# ADVANCES IN CHEMICAL PHYSICS

VOLUME 143

## EDITORIAL BOARD

# ADVANCES IN CHEMICAL PHYSICS

VOLUME 143

*Series Editor*

**STUART A. RICE**

Department of Chemistry
and
The James Franck Institute
The University of Chicago
Chicago, Illinois

A JOHN WILEY & SONS, INC., PUBLICATION

Published by John Wiley & Sons, Inc., Hoboken, New Jersey
Published simultaneously in Canada



***Library of Congress Catalog Number: 58-9935***

ISBN: 978-0-470-50025-5

Printed in the United States of America

10 9 8 7 6 5 4 3 2 1

# CONTRIBUTORS TO VOLUME 143

A. Weiner, Infosys, Mysore, India

Jerome Weiner, Brown University, Division of Engineering, Providence, RI 02912

Thomas Loerting, Institute of Physical Chemistry, University of Innsbruck, Innrain 52a, A-6020 Innsbruck, Austria

Vadim V. Brazhkin, Institute for High Pressure Physics, Troitsk, Moscow Region, 142190 Russia

Tetsuya Morishita, Research Institute for Computational Sciences (RICS), National Institute of Advanced Industrial Science and Technology (AIST), Central 2, 1-1-1 Umezono, Tsukuba, Ibaraki, 305-8568 Japan

Dongping Zhong, Departments of Physics, Chemistry, and Biochemistry, The Ohio State University, Columbus, OH 43210

# INTRODUCTION

Few of us can any longer keep up with the flood of scientific literature, even in specialized subfields. Any attempt to do more and be broadly educated with respect to a large domain of science has the appearance of tilting at windmills. Yet the synthesis of ideas drawn from different subjects into new, powerful, general concepts is as valuable as ever, and the desire to remain educated persists in all scientists. This series, *Advances in Chemical Physics*, is devoted to helping the reader obtain general information about a wide variety of topics in chemical physics, a field that we interpret very broadly. Our intent is to have experts present comprehensive analyses of subjects of interest and to encourage the expression of individual points of view. We hope that this approach to the presentation of an overview of a subject will both stimulate new research and serve as a personalized learning text for beginners in a field.

STUART A. RICE

# CONTENTS

# NONBONDED INTERACTIONS IN RUBBER ELASTICITY

AARON WEINER

*Infosys, Mysore, India*

JEROME WEINER*

*Division of Engineering, Brown University, Providence RI 02912*

## CONTENTS



*Email: Jerome_Weiner@brown.edu

*Advances in Chemical Physics, Volume 143*, edited by Stuart A. Rice

# I. INTRODUCTION

Rubber and rubber-like materials are systems of molecules—monomers or mers—that are subject to two types of interactions. The first type are covalent interactions that tie monomers into long chains, which are typically 100 or more mers long. The second type are nonbonded interactions, which occur between pairs of mers that are not covalently bonded to each other. We are concerned here with an examination of how nonbonded interactions are generally treated in theoretical studies of rubber elasticity and with the limitations of this approach.

## A. Linear Elasticity Theory Notation

Although a key characteristic of the mechanical behavior of rubber-like materials is their ability to undergo large elastic deformations, we will present here some important results from the theory of linear elasticity [1], which is valid only for small deformations. These serve our present purposes better than the nonlinear theory, because of their simpler character and physical transparency.

## B. Stress Tensor

We present first a brief discussion of the stress tensor and the concepts of its mean and deviatoric parts. A rectangular Cartesian coordinate system $(x_i\ i = 1, 2, 3)$ is used throughout. The stress tensor referred to this coordinate system is denoted by $t_{ij}, i, j = 1, 2, 3$. The mean stress is $\Pi = t_{ii}/3 = (t_{11} + t_{22} + t_{33})/3$, where the summation convention (repeated indices imply sum over $i = 1, 2, 3$) is used here and throughout. Positive stress components are tensile. The pressure $p = -t_{ii}/3$ so that positive pressure is compressive. The deviatoric stress ${}^{D}t_{ij}$ is defined as

$$ {}^{D}t_{ij} = t_{ij} - t_{kk}\delta_{ij}/3 \tag{1} $$

where $\delta_{ij}$ is the Kronecker delta, $\delta_{ij} = 1, i = j$, and $\delta_{ij} = 0, i \neq j$.

## C. Small Strain Tensors

We next introduce the small strain tensor $e_{ij}$ and its deviatoric part

$$ {}^{\mathrm{D}}e_{ij} = e_{ij} - e_{kk}\delta_{ij}/3. $$

Consider next the system of volume v in a state of stress $t_{ij}$ and subject to a small change in strain at a rate of $\dot{e}_{ij}$. (the superimposed dot denotes the derivative with respect to time). Then the rate of doing work $\dot{W}$ on the system is

$$ \dot{W} = \mathrm{v} t_{ij}\dot{e}_{ij} = \mathrm{v}\, {}^{D}t_{ij}\, {}^{D}\dot{e}_{ij} + p\dot{\mathrm{v}} \tag{2} $$

where the relation

$$ \dot{e}_{kk} = \dot{\mathrm{v}}/\mathrm{v} \tag{3} $$

has been used. This relation thus provides the picture that ${}^{D}t_{ij}$, which is the deviatoric part of the stress tensor, does work because of the part of the deformation that corresponds to ${}^{D}e_{ij}$, which is a change in shape at constant volume, whereas the mean stress (or pressure) does work because of a change in volume at constant shape. The stress strain relations for an isotropic elastic solid can be written in the form $t_{ij} = 2G\ {}^{D}e_{ij}, p = Ke_{kk} = K\Delta \mathrm{v}/\mathrm{v}$, where $G$ is the shear modulus and $K$ is the bulk modulus. For a rubber elastic solid (above the glass transition), $K$ is much larger than $G$ (typically $K/G \approx 1000$); i.e., it is much easier to change the shape of a rubbery solid than its volume. The large ratio $K/G$ is sometimes cited as an indication that the mechanisms of generation of pressure and deviatric stress are completely different. This may be so in terms of a molecular or chain theory of rubber elasticity, but in terms of atomic interactions, we will observe [Eqs. (27) and (28)] that they are, in fact, closely related. As a result, in some theoretical discussions [2], rubber is idealized as a constant volume solid and is characterized by only a single elastic constant, $G$. The goal of a theory (molecular or atomic) is then regarded as restricted to the determination of the deviatoric stress ${}^{D}t_{ij}$ because of a prescribed ${}^{D}e_{ij}$ ; the mean stress or pressure $p$ is regarded as given by prescribed boundary conditions. Thus, in the frequently discussed experiment of a constant volume extension in the $x_1$ direction, the stress components are $t_{11}$, and by cylindrical symmetry, $t_{22} = t_{33}$. Then, ${}^{D}t_{11} = t_{11} - (t_{11} + t_{22} + t_{33})/3 = 2(t_{11} - t_{22})/3$. The restriction of the analysis to ${}^{D}t_{11}$ is sometimes stated as a restriction to stress differences because $p$, the pressure, is not specified. Note, however, that the equivalence of a deviatoric stress component to a simple stress difference is only possible in a state of stress with two equal principal stresses, i.e., a cylindrical state of stress.

Because of the large ratio $K/G$, most studies of rubber elastic behavior are restricted to constant volume deformations, i.e., to ${}^{D}e_{ij}$, and to the resulting deviatoric stress ${}^{D}t_{ij}$. Although this restriction is widely made, it is often tacit. Because of its importance, we make it explicit here because it has important consequences for the current discussion.

### D. Nonbonded Atomic Interactions

The early molecular theories of rubber elasticity were based on models of networks of long chains in molecules, each acting as an entropic spring. That is, because the configurational entropy of a chain increased as the distance between the atoms decreased, an external force was necessary to prevent its collapse. It was understood that collapse of the network to zero volume in the absence of an externally applied stress was prevented by repulsive excluded volume (EV) interactions. The term "nonbonded interactions" was applied to those between atom pairs that were not neighboring atoms along a chain and interacting via a covalent bond.

In the usual development of the theory, the important assumption was made that the nonbonded interactions, although certainly present, contributed only to the mean stress $p$ and made zero contribution to the deviatoric stress ${}^{D}t_{ij}$. Because as noted, the earlier restricted theories of rubber elasticity were concerned only with understanding the deviatoric stress and not the pressure, it then seemed logical to neglect the nonbonded interactions in theory development.

The growth in the 1950s of the availability of digital computers for research led to the development of the method of molecular dynamics (MD) as a tool for gaining insight into the behavior of various systems on the atomic and molecular levels [3]. In the 1970s, models of polymeric systems began to be studied in this way. A series of simulations by Gao and Weiner [4] were directed to the examination of the basic assumption that nonbonded interactions did not contribute to the deviatoric stress ${}^{D}t_{ij}$. The model system consisted of a collection of atoms that formed chains in a periodic unit cell with volume v. (In discussions of this idealized model system, we use the terms "mer" and "atom" interchangeably.) The atoms interact through the following two types of two-body potential: $u_b$, which represents the covalent bonds responsible for the chains, and $u_{nb}$, which represents the interaction between any pair of nonbonded atoms. Covalent bonds are approximated with a potential

$$u_b(r) = \frac{1}{2}\kappa(r - a)^2 \tag{4}$$

where $r$ is the distance between adjacent atoms on a given chain and $a$ is the bond length. The interaction between nonbonded atoms is modeled using a truncated Lennard-Jones potential.

$$u_{nb}(r) = \begin{cases} 4 \in_{LJ} \left[\left(\frac{\sigma_{LJ}}{r}\right)^{12} - \left(\frac{\sigma_{LJ}}{r}\right)^{6}\right] & r \le R_c \\ u(R_c), & r > R_c \end{cases} \tag{5}$$

Parameter values were chosen so that $u_b$ models a stiff covalent bond, whereas the repulsive portion of $u_{nb}$ approximates a hard sphere potential of diameter $\sigma_{LJ}$, which was set equal to the bond length $a$ so that the chain becomes the familiar "pearl necklace" model.

The assemblage of chains is constructed to represent the affine network model of rubber elasticity in which all network junction positions are subject to the same affine transformation that characterizes the macroscopic deformation. In the affine network model, junction fluctuations are not permitted so the model is simply equivalent to a set of chains whose end-to-end vectors are subject to the same affine transformation. All atoms are subject to nonbonded interactions; in the absence of these interactions, the stress response of this model is the same as that of the ideal affine network.

To construct the model, $N_c$ distinct chains with $N_a$ atoms each were generated by independent random walk routines; this procedure led to a macroscopically initially isotropic system. The velocities of the end atoms of each chain were constrained to be equal, so that the chain vectors remained constant, The motive for permitting the chains to translate in this fashion is to permit the sampling of different neighborhoods of interaction. The end pairs, as well as the remaining atoms, were set in random motion corresponding to temperature $T$. The end-to-end vectors of the chains as well as the basic periodic cell (used in the MD simulation) were subjected to an affine deformation corresponding to a constant volume elongation $\lambda$ in the $x_1$ direction. This model is an idealization. Its chain vectors are subjected by fiat to the prescribed affine deformation; they are not part of a system-spanning network that serves to transfer the imposed deformation from the system boundary to its interior. Nevertheless, it serves to represent the effect of nonbonded interactions on the affine network model, and its stress–strain relation reduces to that predicted by that model when nonbonded interactions are absent. With the end-to-end chain vectors controlled in this fashion, we refer to this system as in the network mode. It may be modified to represent a polymer melt with the only change being that in the melt mode, the end atoms of the chains are all also free to engage in unconstrained thermal motion.

After a sufficiently long run in the network mode, the equilibrium value of the stress tensor $t_{ij}$ was determined by application of the virial formula [5, 6]

$$\mathrm{v}t_{ij} = \sum_{\alpha} \langle f_i^{\alpha} r_j^{\alpha} \rangle \tag{6}$$

where the sum is over all atomic pairs $\alpha$ with separation vector $r_j^{\alpha}$ and interacting force $f_i^{\alpha}$. Brackets denote a time average.

An important parameter in this series of simulations is the packing fraction $\eta$ of the atoms; for $n$ spheres of diameter $a$ confined to a volume v,

$$\eta = \pi n a^3/(6\mathrm{v}) \tag{7}$$

An equivalent parameter used in these discussions is the reduced density

$$\rho^* = na^3/\mathrm{v} = 6\eta/\pi \tag{8}$$

The initial series of simulations was performed for $0 < \rho^* < 1$. Because the purpose of these simulations was to test the common assumption that the nonbonded interactions did not contribute to the deviatoric stress, the sum in Eq. (6) was separated as in Eq. (9) to show the bonded and nonbonded interactions explicitly. The first sum is over all atom pairs that interact through $u_b$, the second is over all pairs that interact through $u_{nb}$.

$$\mathrm{v}t_{ij} = \sum_{\alpha \in b} \frac{\partial u_b^{\alpha}}{\partial r_i} r_j + \sum_{\alpha \in nb} \frac{\partial u_{nb}^{\alpha}}{\partial r_i} r_j^{\alpha} \tag{9}$$

For low values of $0 < \rho^* < 0.3$, they seemed to support the assumption that nonbonded interactions made only an isotropic contribution to the stress. However, as $\rho^*$ increased, the anisotropic or deviatoric contribution increased and became substantial.

An examination of the nature of a typical nonbonded interaction with a generic atom revealed that the inherently isotropic two-body interaction of Eq. (5) was rendered anisotropic by steric shielding from the neighboring atoms surrounding the generic atom, provided this neighborhood had been rendered statistically anisotropic by the applied deformation. Simulations described in Section II.C and in Ref. [7] show that the anisotropy in the neighborhood of a generic atom decays rapidly with distance from that atom. It then follows from Eq. (5) that the nonbonded contribution to the deviatoric stress is primarily caused by the short range, approximately hard sphere, behaviors of $u_{nb}(r)$.

### E. Monomer Packing Fraction of Rubber

Data for the monomer packing fraction for various polymeric systems that are rubber-like at standard conditions are provided by van Krevelen [8]. There, the monomer is regarded as an impenetrable object of geometric shape composed of spheres of radii that correspond to the atomic radii of the constituent elements arranged in a known configuration with known bond lengths. It is then possible by geometric calculation to compute the monomer volume $\mathrm{v}_m$ and the Van der Waals molar volume $V_W = N_{\mathrm{Av}}\mathrm{v}_m$. The molar volume of the rubber $V_r = N_{\mathrm{Av}}M/\tilde{\rho}$, where $M$ is the monomer molecular weight and $\tilde{\rho}$ is the rubber mass density. Although $V_r$ and $V_W$ each vary widely for a group of 35 different systems ($24 < V_r < 240$; $15 < V_W < 150$), both in cm$^3$/mol) the packing fraction $V_W/V_r = 0.625$ with little variation. (Actually, van Krevelen reports for the data the common ratio $V_r/V_W = 1.6 \pm 0.035$ and does not use the packing fraction concept or terminology. This may partly account for the relative absence of this surprising and important result from the rubber elasticity literature.)

The universal packing fraction, $\eta = 0.625$, for the mers of rubber-like polymer systems corresponds to the random close packing of hard spheres. The existence of this universal value may be motivated as follows: Assume first the absence of nonbonded interactions and consider a network of Gaussian chains $\gamma$ with chain vectors $\mathbf{R}(\gamma)$ occupying a volume v. The force $\mathbf{f}(\gamma)$ required to maintain the chain vector fixed at $\mathbf{R}(\gamma)$ is

$$f_i(\gamma) = 3kTR_i(\gamma)/\langle R^2 \rangle_0 \qquad (10)$$

where $R = |\mathbf{R}|$ and $\langle R^2 \rangle_0$ is the mean-square end-to-end distance of the corresponding free chain.

The stress in the network can then be expressed by the virial theorem, Eq. (6), where in the molecular theory, the set of particles considered is not all the atoms of the system but only the end atoms of each chain, and they are regarded as subject to the force in the chain connecting them. That is, the stress $t_{ij}$ is then given by

$$t_{ij} = \frac{1}{V} \sum_{\gamma} f_i(\gamma) R_j(\gamma) \tag{11}$$

Then the pressure $p$ is given by

$$p = -t_{ii}/3 = -\frac{kT}{V} \sum_{\gamma} R^2(\gamma)/\langle R^2 \rangle_0 \tag{12}$$

which is always negative for nonzero $\mathbf{R}(\gamma)$. Therefore, in the absence of a balancing positive pressure, the system of chains must collapse to a point. This balancing pressure is provided by the repulsive EV mer–mer interaction, which comes into existence only when the mers approach the random close-packed state; the asymptotic behavior of this balancing pressure for a system of $n$ mers may be written [9] as

$$p = 3nkT/(1 - \eta/0.63) \tag{13}$$

The resulting physical picture of a rubber-like system as a close-packed collection of mers is radically different from the two-phase image introduced by James and Guth [10]. The latter represents rubber as a network of chains, which act as entropic springs in tension, embedded in a bath of simple liquid. The bath gives rise to an isotropic pressure, whereas the network is responsible for the deviatoric stress. More recent physical pictures consider as well the distribution of network junctions in the liquid and the action of these junctions as constraints on the free motion of a generic chain of the network. The current description is on the mer or atomic level and treats the full stress tensor, both the mean and deviatoric portions, in terms of atomic interactions.

### F. Simulation Results

We present here some computer simulation results for the model described in Section I.E that illustrate the important role of monomer packing fraction $\eta$, Eq. (7), or equivalently of the reduced density $\rho^*$, Eq. (8), on the response of the model to a constant volume extension $\lambda$. In these simulations, $\lambda = 1.6$. Simulations were run for $T^* = kT/\epsilon_{LJ}$ in the range $1.4 < T^* < 5$, and for $\rho^*$ in the range $0.8 < \rho^* < 1.2$. The system consists of $N_c = 40$ chains, each with

$N_a = 20$ atoms. The reduced density is varied by setting the volume v of the basic cell according to the following equation:

$$\rho^* = \frac{na^3}{\mathrm{v}} = \frac{N_c N_a a^3}{\mathrm{v}} \tag{14}$$

The simulation results for $t_{11} - t_{22}$ are reported in terms of the nondimensional

$$\sigma = (t_{11} - t_{22})\mathrm{v}/kT \tag{15}$$

These simulation results, including the effects of nonbonded interaction, may be compared with $\sigma_{\text{ideal}}$, with these interactions absent,

$$\sigma_{\text{ideal}} = N_c(\lambda^2 - \lambda^{-1}) \tag{16}$$

Results are shown in Fig. 1. It is observed that the behavior changes sharply at $\rho^* = 1.0$. We may understand why this value is critical as follows: If v is considered divisible into a lattice of $r$ cubes of side $a$, so that $\mathrm{v} = ra^3$ and $\rho^* = n/r$, then $\rho^* = 1$ permits one sphere of diameter $a$ to be placed in each of the cubes.

As a preliminary to the analysis of these results, we first recall the thermodynamic relation

$$\frac{\partial U}{\partial \lambda} = f - T\frac{\partial f}{\partial T} \tag{17}$$

where $f$ is the force exerted in one-dimensional extension and $U$ is the internal energy. This relation has the useful graphical interpretation shown in Fig. 2; by extrapolating the force-temperature relation to zero temperature, the stress may be divided into an energetic and an entropic component.

We now return to the results of Fig. 1, which displays a set of curves $\sigma(\rho^*, T^*)$ for discrete values of $T^*$ in the range $1 < T^* \leq 5$. For certain intervals, the curves for different $T^*$ become identical, namely

$$\sigma(\rho^*, T^*) = F(\rho^*) \quad T^* > 1.4 \quad 0.8 < \rho^* < 1$$
$$\text{and } \sigma(\rho^*, T^*) = F(\rho^*) \quad 3.6 < T^* < 5 \quad 0.8 < \rho^* < 1.2.$$

For each of these intervals, it follows from the definition of $\sigma$, Eq. (15), that

$$T^*\sigma = (t_{11} - t_{22})\mathrm{v}/\in_{LJ} = \mathrm{T}^*\mathrm{F}(\rho^*) \tag{18}$$

and we conclude from the thermodynamic analysis of Eq. (17) and its graphical representation, Fig. 2, that the model system behaves purely entropically in each of these regions. These results are illustrated in Fig. 3, by replotting the data from

Fig. 1 for $\rho^* = 0.8$ and 1.2 [20]. We summarize the results of these simulations as follows:

i. For $\rho^* = 0.8$, the system is entropic for all $T^* > 0$, but the system with $\rho^* = 1.2$ is purely entropic only for $T^* > 3$ ; from Figs. 2 and 3, we observe that for $T^* < 3$, it acquires an increasing internal energy component. We interpret the latter as representing a glass transition beginning at $T^* \approx 3$.

ii. If we postulate a stress-stretch relation in the entropic regime of the form $t_{11} - t_{22} = G(\rho^*)(\lambda^2 - \lambda^{-1})$, then simulations indicate $G(1.2)/G(0.8) \approx 2.3$.

## G. Glass Transition

We emphasize here two important results that follow from simulations of this model and from the examination of Figs. 1, 2, and 3.

(a) From Fig. 1, we conclude that for this model, no glass transition occurs for $0.8 < \rho^* < 1$.

(b) From Figs. 1–3, we conclude that such a transition exists for $\rho^* > 1$.

If we assume that the general character of these figures is model independent, these results add support to van Krevelen's data compilation because a glass

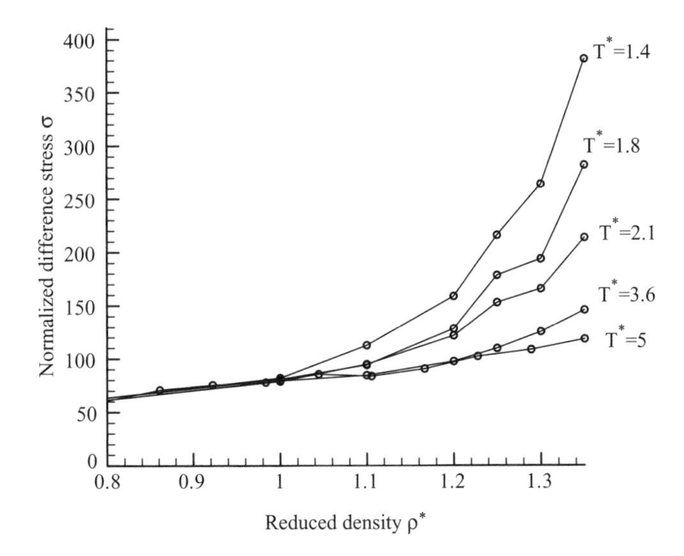


**Figure 1.** Normalized difference stress as a function of reduced density, for several values of normalized temperature, for stretch $\lambda = 1.6$.

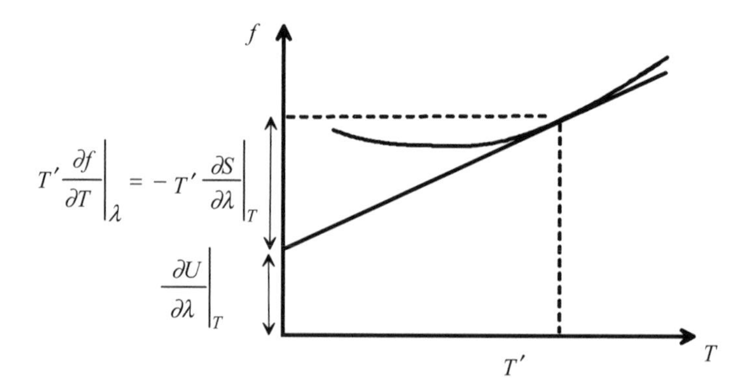


**Figure 2.** A schematic based on Eq. (17) illustrating the procedure used to identify energetic and entropic contributions to the total difference stress plotted in Fig. 3.

transition at sufficiently low $T$ is a universal characteristic of rubber-like systems.

Because of the high values of $\rho^*$ considered, it is difficult to generate allowable initial conditions for MD simulations; the methods of doing so are described in detail in Ref. [11]. By following these procedures, 250 independent ensembles are generated, and the results presented are time averages over these. The question of the rate of temperature change in experiments to obtain glass transition temperatures is bypassed in this way and not considered here.

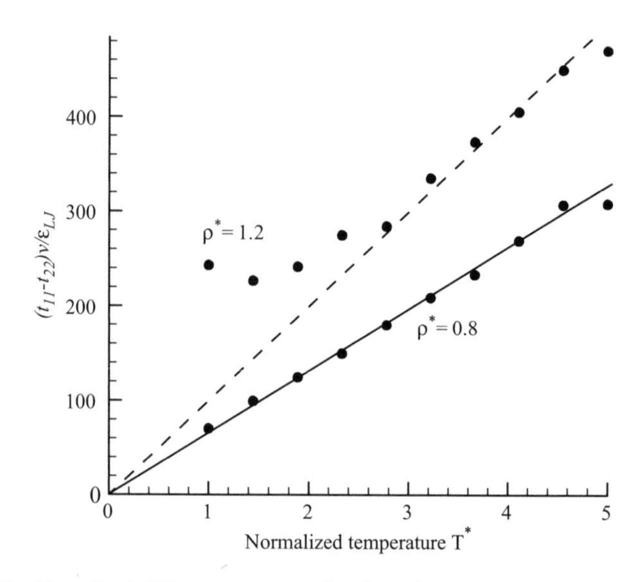


**Figure 3.** Normalized difference stress as a function of temperature for stretch $\lambda = 1.6$.

## II. CONCEPTS AND FORMULATIONS FOR DENSE INTERACTING SYSTEMS

In Section I, the discussion dealt with the significant role of nonbonded interactions in the development of the full stress tensor, mean plus deviatoric, in rubber elasticity, in the important high reduced density regime $\rho^* > 1$. Here, we present some concepts and formulations that apply to this regime.

### A. Chain Force

The concept of a long chain molecule acting as an entropic spring plays a central role in most molecular theories of rubber elasticity. To what extent does this concept remain valid and useful in dense systems of interacting chains? This question has been considered by MD simulation in Ref. [12].

Consider first an isolated, ideal chain (no nonbonded interactions) in Fig. 4, with its two end atoms fixed at a distance $r$ apart in the $x_1$ direction, with the remaining chain atoms in thermal motion. Each end atom is subject to the pull exerted on it by the adjacent covalent bond, and to keep the end atoms at rest, external forces as

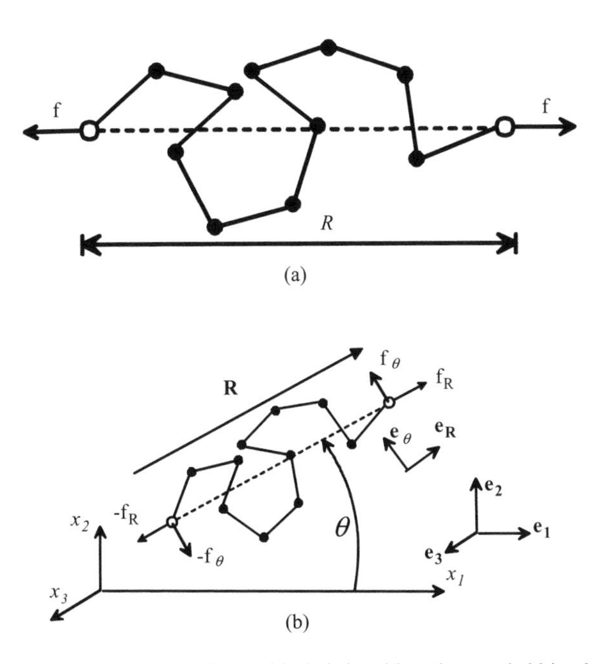


**Figure 4.** (a): Schematic illustrating an ideal chain with end atoms held in place by external forces. (b) Schematic representation of the force acting on a chain in a stretched network when the chain vector **R** makes an angle $\theta$ with the stretch axis $x_1$. Shown are $f_R$ and $f_\theta$, the axial and transverse components of the chain force **f**.

shown must be applied. As a particular example, consider a freely jointed isolated chain with $n$ bonds, each of length $a$, and fixed end-to-end distance $r$. Then, from Eq. (10) with $\langle R^2 \rangle_0 = na^2$, the Gaussian attractive force $f$ is

$$\frac{fa}{kT} = \frac{3r}{na} \tag{19}$$

Now consider this same chain immersed in a melt of like chains; all atoms are in thermal motion except the end atoms of the tethered chain, and in addition to the bonded interactions, all atom pairs are subjected to nonbonded interactions through $u_{nb}$, Eq. (5).

Now, in addition to the forces from their bonded adjacent atoms, the end atoms are subject also to nonbonded interactions from those atoms in the surrounding melt as well as possibly those from other atoms in the chain. Again, it is necessary to apply external forces to the end atoms of the tethered chain to keep them in equilibrium.

This example has been considered in detail in Ref. [12]. When the surrounding melt, was isotropic as in a free unstressed melt, it was found that the required external forces are in the same directions as those shown in Fig. 4a. However, the end atoms are not being pulled inward by the chain but are being pushed inward by the external nonbonded interactions. Therefore, it is no longer correct to speak of the chain as a spring in tension,although the required end forces make it seem so. The mechanism whereby the chain end atoms are pushed inward by external nonbonded interactions is related to the simpler case of a hard-sphere gas [13]. There, pairs of atoms, say atom A and B, seem to be attracted to each other at close range because atom A is shielded from further external interactions coming from the direction of its close neighbor B, The origin of the apparent chain force on a polymer chain caused by nonbonded interactions also involves directional shielding but is more complex than the simple hard-sphere case. Here, the chain end atoms are subject to the following three classes of interactions: (1) covalent from the chain atoms adjacent to the end atoms, (2) intrachain nonbonded atoms, and (3) interchain nonbonded atoms. The relative values of those contributions depend on the parameters $r/na$ for a chain of $n$ bonds of length $a$ with ends fixed at a distance $r$ apart, and on $\rho^*$, the reduced density of the surrounding chains.

We have computed the apparent attractive force on the end atoms of an immersed tethered chain by an MD simulation of the system. We refer to this force as the chain force to distinguish it from the Gaussian attractive force on the corresponding isolated chain. Results as a function of $\rho^*$ and $T^*$ of the surrounding isotropic melt are shown in Figs. 5 and 6.

Consider next the case in which the surrounding melt is replaced by the same set of chains in the network mode, that is, with the chain vectors controlled corresponding to an applied deformation, which is a constant volume extension $\lambda$ in the same direction as the tethered chain. Then, as shown in MD simulations,

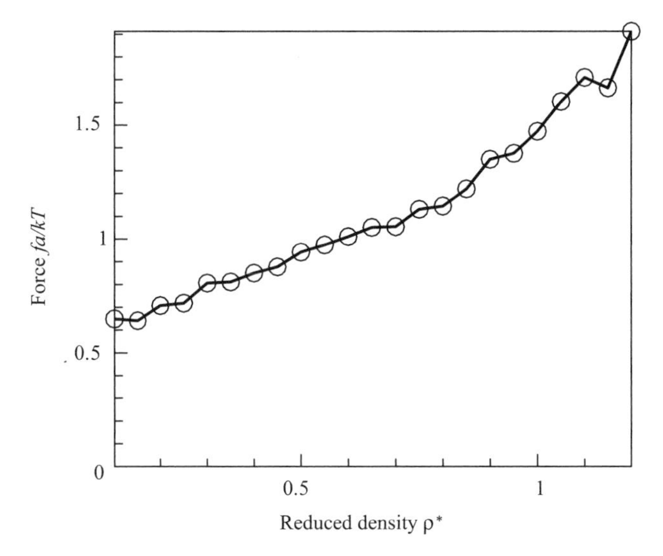


**Figure 5.** The results of computer simulations of the variation of the average end force $f$ acting on the chains in a pearl necklace network. The system consisted of 20 identical chains, each with $N_b = 14$ bonds of length $a$, 13 free with one tethered. For the latter, the fixed end-to-end distance $r = 6.62a$. The reduced temperature was $T^* = 4$.

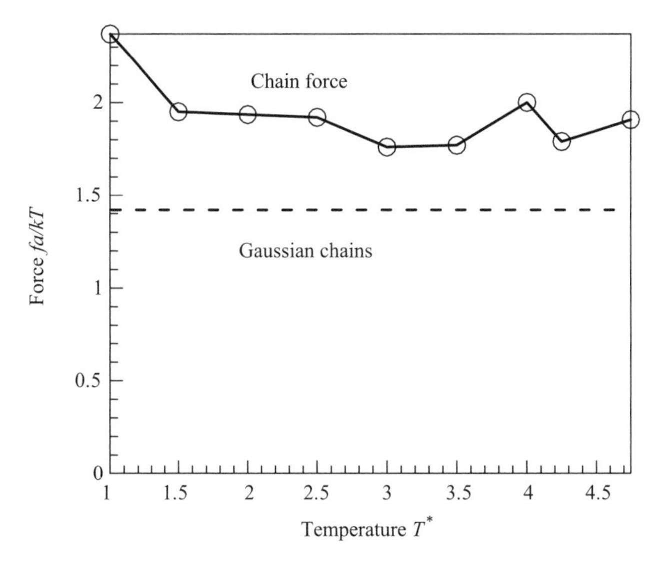


**Figure 6.** Conditions as in Fig. 5, which show dependence of $fa/kT$ upon $T^*$. Value of $fa/kT^* = 1.42$ for Gaussian chains from Eq. (19).

the effective chain force is no longer coaxial with the tethered chain vector, as in Fig. 4a, but it may be represented by a coaxial vector plus an applied couple, as in Fig. 4b. The strength of this couple increases with $\lambda$. When a network model of chains that obey this coaxial plus couple force law is evaluated by the virial theorem, the couples are found to have a softening effect. Additional details can be found in Ref. [12].

### B. Nonbonded Interactions: Integral Formulation

Consider a generic atom $\beta$ in the rubber model of Section I. From the virial Eq. (6), its nondimensional contribution $\Delta\sigma_{ij}^{nb}(\beta)$ to the stress caused by the nonbonded interaction through the potential $u_{nb}(r)$ is

$$\sigma_{ij}^{nb}(\beta) == \frac{1}{2kT}\sum_{\alpha(\beta)} f_i^{\alpha} r_j^{\alpha} = \frac{1}{2kT}\sum_{\substack{\alpha(\beta)\\ \alpha\in nb}} (u'_{nb}(r_\alpha) r_i^{\alpha} r_j^{\alpha}/r_\alpha) \tag{20}$$

where the sum is over all atom pairs $\alpha(\beta)$ that interact through $u_{nb}(r)$ with $\beta$; $r_j^{\alpha}$ are components of the connecting vector of pair $\alpha(\beta)$, and $f_i^{\alpha} = \frac{\partial u_{nb}}{\partial r_i^{\alpha}} = u'_{nb}(r_\alpha) r_i^{\alpha}/r_\alpha$ with $r_\alpha = \sqrt{r_i^{\alpha} r_i^{\alpha}}$.

The total nonbonded contribution to the stress $t_{ij}$ is then $\sigma_{ij}^{nb} = \sum_\beta \sigma_{ij}^{nb}(\beta)$, which is the sum of $\sigma_{ij}^{nb}(\beta)$ over all atoms $\beta$ that engage in nonbonded interactions. This sum may be written in an integral form by use of the radial distribution function $g(r)$, where

$$dn = \rho g(r) 4\pi r^2 dr \tag{21}$$

where $\rho = n/\mathrm{v}$ is the atom number density and $dn$ is the number of atoms in a spherical shell between radii $r$ and $r + dr$ surrounding a representative atom in the system and in nonbonded interaction with it. The nonbonded contribution to the stress can be expressed in terms of $g(r)$ as

$$\sigma_{ij}^{nb} = \frac{2\pi\rho^2 \mathrm{v}}{kT}\int_0^{\infty} u'_{nb}(r) g(r) S_{ij}(r) r^3 dr \tag{22}$$

where

$$S_{ij}(r) = \left\langle \frac{r_i r_j}{r^2} \right\rangle \tag{23}$$

with the average taken over all atoms in the shell $r$ to $r + dr$. The nondimensional nonbonded contribution ${}^D\sigma_{ij}^{nb}$ to the deviatoric stress ${}^D t_{ij}$ is

$${}^D\sigma_{ij}^{nb} = \frac{2\pi\rho^2 \mathrm{v}}{kT}\int_0^{\infty} u'_{nb}(r) g(r) {}^D S_{ij}(r) r^3 dr \tag{24}$$

whereas the nonbonded contribution to the dimensionless mean stress is

$$\Pi_{nb} = \frac{2\pi\rho^2 \mathrm{v}}{3kT} \int_0^\infty u'_{nb}(r) g(r) r^3 dr \tag{25}$$

because $S_{ii}(r) = 1$.

This formulation shows clearly that the deviatoric part of $\sigma_{ij}^{nb}$ depends on the anisotropy of the atomic distribution about any generic atom $\beta$ because ${}^D S_{ij} = 0$ for an isotropic distribution. It has been shown [7] that the anisotropy of the distribution is localized and therefore that ${}^D\sigma_{ij}^{nb}$ is, to an overwhelming extent, caused by the repulsive excluded volume interaction. See Section II.C and Fig. 7.

### *1. Hard Sphere Limit*

In this section, we develop the formulation of the nonbonded interactions in the hard sphere limit. Although the simulations are based on the repulsive part of $u_{nb}(r)$ as defined in Eq. (5), the hard sphere formulation permits a simpler physical interpretation that is valuable.

By the same mathematical technique used in the theory of simple fluids[1] the hard-sphere limit for Eq. (24)

$${}^D\sigma_{ij}^{nb} = -2\pi\rho^2 \mathrm{v} d^3 g(d)\, {}^D S_{ij}(d) \tag{26}$$

where $d$ is the hard-sphere diameter

We also confine attention here to $\Pi_{nbr}$, that portion of $\Pi_{nb}$ caused by the repulsive part of $u_{nb}$, and the hard-sphere approximation to Eq. (25) leads to

$$\Pi_{nbr} = \frac{-2\pi\rho^2 \mathrm{v} d^3}{3} g(d) \tag{27}$$

A comparison of Eqs. (26) and (27) written for a generic atom shows

$${}^D\sigma_{ij}^{nb}(\beta) = 3\Pi_{nbr}(\beta)\, {}^D S_{ij}(d) \tag{28}$$

[1]J.P. Hansen and I.R. McDonald [14]. Briefly, the argument is as follows. Let $A = \frac{1}{2}\int_0^\infty u'(r) f(r) dr$, where $f(r)$ is an arbitrary continuous function, and set $y(r) = e^{\frac{u(r)}{kT}}$. Then $A = \frac{1}{kT}\int_0^\infty u'(r)$ $y(r) e^{\frac{-u(r)}{kT}} f(r) dr = -\int_0^\infty y(r) f(r) \frac{d}{dr} e^{\frac{-u(r)}{kT}} dr$. In the hard-sphere limit of $u(r)$, $\lim e^{\frac{-u(r)}{kT}} = \begin{matrix} 0 \; for \; r < d \\ 1 \; for \; r \geq d \end{matrix}$ and $\lim \frac{d}{dr} e^{\frac{-u(r)}{kT}} = \delta(r - d)$, the Dirac delta function, so that $A = -\int_0^\infty y(r) f(r) \delta(r - d) dr = -\lim_{r \to d+} y(r) f(r) = -f(d)$.

Equations (27) and (28) have a simple and important physical interpretation. The contribution $\Pi_{nbr}(\beta)$ that atom $\beta$ makes to the system pressure caused by hard-sphere impacts from other atoms does not, as observed from Eq. (27), depend on the distribution of these impacts, over the sphere of atom $\beta$.[2] Its contribution ${}^{D}\sigma_{ij}^{nb}(\beta)$ to the deviatoric stress, however, does depend on the anisotropy and is measured by the shield tensor ${}^{D}S_{ij}(d)$.

To make the shield tensor ${}^{D}S_{ij}(d)$ explicit, we introduce spherical coordinate systems $(r, \theta, \varphi)$ with the polar axis in the $x_i$ direction, $i = 1, 2, 3$ (Fig. 1). It then follows from Eq. (23) that ${}^{D}S_{ij}(d) - S_{ij}(d) - \frac{1}{3}\delta_{ij}S_{rr}$ has the values

$$ {}^{D}S_{ii}(\text{no sum}) = \frac{2}{3}\langle P_2(\theta_i)\rangle \tag{29} $$

where $P_2(\theta_i) = \frac{1}{2}(3\cos^2\theta_i - 1)$ and $i \neq j$,

$$ {}^{D}S_{ij} = \langle \cos\theta_i \sin\theta_i \cos\varphi_j \rangle = \left\langle \frac{1}{2}\sin 2\theta_i \cos\varphi_j \right\rangle \tag{30} $$

where the average is taken over the impact sphere,

### C. Nature of Nonbonded Interactions

The potential $u_{nb}(r)$ for nonbonded interactions has both repulsive and attractive regimes. We have performed an MD simulation [7] of our model to determine the relative importance of those regimes. The simulation involves first a loading period in which the system is subjected to a constant volume extension $\lambda$ in the $x_1$ direction. During this period, the positions of the chain end atoms are controlled to apply the deformation, whereas the remaining atoms are free to engage in thermal motion. At the end of this period, when the system is in equilibrium in the deformed state, the chain end atoms are also freed to engage in thermal motion; in this unloading period, the deviatoric stress relaxes ultimately to zero.

The deformed system has cylindrical symmetry, with $t_{22} = t_{33}$ so that ${}^{D}t_{11} = \frac{2}{3}(t_{11} - t_{22})$. During the unloading period, the range of integration in Eq. (24) is subdivided $q\delta \leq r \leq (q+1)\delta, q = 0, 1, \ldots, n$ with $n$ a convenient

[2] More generally this may be seen from Eq. (6),

$$ \sigma_{jj}(\beta) = 3\Pi(\beta) = \sum_{\alpha(\beta)} \langle f_j^{\alpha} r_j^{\alpha} \rangle $$

so that the contribution of $\beta$ to $\Pi$ involves only a sum of radial components of interaction forces, not their distribution.

cutoff, and the program computes $\Delta\sigma_{nb}(\delta; r, t)$, the contribution, to

$$\sigma_{nb} = \frac{\text{V}}{kT}(t_{11} - t_{22})(t) \tag{31}$$

made by the atoms in the radial interval $\delta$ with midpoint $r$ at time $t$ (measured from start of unloading), included in the simulations was the case of the simple liquid system, i.e., with $u_b = 0$. An unexpected result is that the normalized quantity

$$\Delta\sigma_{nb}(\delta; r, t)/\sigma_{nb}(t) \tag{32}$$

is time independent. The results are shown in Figures 7(a), 7(b) and 7(c).

Some conclusions that can be drawn from this simulation of stress relaxation in an atomic model melt are as follows:

(1) The significant nonbonded contribution to deviatoric stress in the melt is made by the strong, short-range, repulsive excluded volume interaction.
(2) The deviatoric stress contribution made by a generic atom is caused by the anisotropy of the atomic distribution in its neighborhood. As the anisotropy is localized, the attractive long-range interaction has only a minor effect.
(3) From the close relation of the atomic model employed when acting either as a melt or as a network, this conclusion applies as well in the latter case.

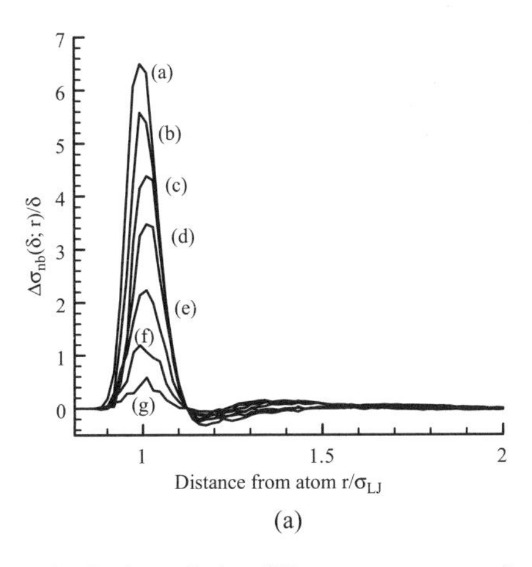


(a)

**Figure 7(a).** Radial distribution of the difference stress contributed by Lennard-Jones interactions. Results are for $N_b = 11$, $\rho^* = 1$, and are shown for normalized times $t/t_0 =$ (a) 0.02; (b) 0.04; (c) 0.08; (d) 0.2; (e) 0.4; (f) 0.8; (g) 2.0. $N_b$ is number of bonds per chain. $\sigma_{LJ}$ is Lennard-Jones parameter, Eq. (5).

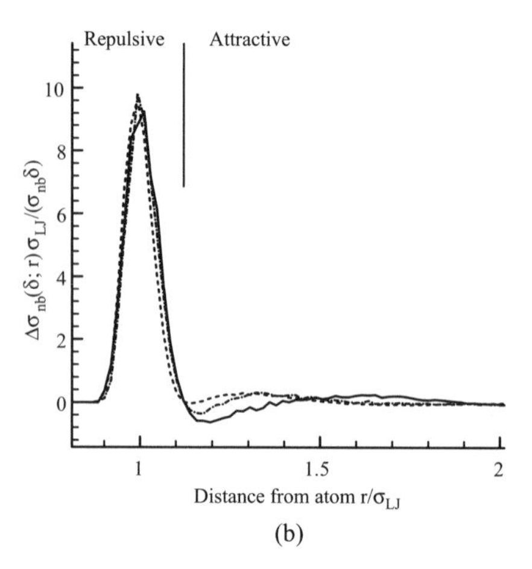


**Figure 7(b).** Normalized radial distribution of difference stress contributed by Lennard-Jones interactions. Results are for $N_b = 11$, $\rho^* = 1$, and are shown for normalized times $t/t_0 = 0.02$ (dashed line), 0.2 (chain line), and 2 (solid line).

(4) The time independence of the normalized stress contribution, Eq. (32), together with the integral formulation for ${}^D\sigma_{ij}^{nb}$, Eq. (24), indicates that in the atomic model, stress relaxation takes place through the relaxation of the local atomic distribution anisotropy as measured by ${}^D S_{ij}$.

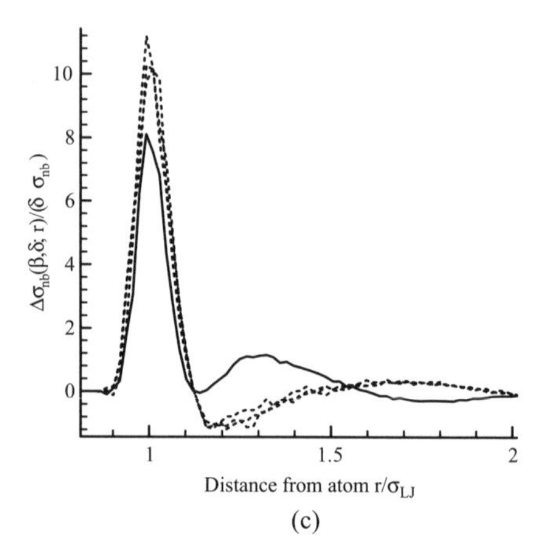


**Figure 7(c).** Normalized steady state radial distribution of difference stress contributed by Lennard-Jones interactions, for a simple liquid $N_b = 0$ (solid line) and polymer melts with $N_b = 1, 3$ and 11 (dashed lines). Reduced density $\rho^* = 0.9$ for all cases. The results for the polymer melts are indistinguishable to within the statistical scatter of the simulations.

### D. Bonded Interactions

We next consider the contributions to the deviatoric stresses ${}^{D}t_{ij}$, which are made by the chain bonds by the spring potential $u_b(r)$, in the MD simulation of our atomic model corresponding to a network (not a melt) under extension $\lambda = 2$ in the $x_1$ direction. The average force $f$ in a covalent bond, modeled by Eq. (4), is

$$f = \kappa\langle r - a\rangle \tag{33}$$

In the ideal case, $\rho^* = 0$, $f \simeq 2kT/a$. This may be interpreted as the tensile centrifugal force caused by the thermal motion of the bond. As the value of $\rho^*$ of the system increases, the value of the forces $f$ decreases and, for larger values of $\rho^*$, becomes negative (compressive). This change is caused by the phenomenon of steric shielding of the EV interactions, as shown schematically in Fig. 8 for the interaction (collision) of nonbonded atoms pair $\alpha_3$ and $\beta$. Because of the steric shielding of $\beta$ by the bonded pair $\alpha_1$ and $\alpha_2$, the interaction of $\alpha_3$ and $\beta$ will occur more readily in configuration (a), when it puts the bonds into compression than in configuration (b). Thus, the bond forces at moderate $\rho^*$ make a negative contribution to the difference stress $t_{11} - t_{22}$, whereas the nonbonded interactions make a positive contribution, as shown in Fig. 9. Also shown is the ideal solution, Eq. (16). It is observed that at low values of $\rho^*$, the simulation results are in reasonably good agreement with the ideal solution, but as $\rho^*$ increases toward $\rho^* = 1.2$, the simulation value that includes both nonbonded and bonded interactions increases by about 30% of the ideal value.

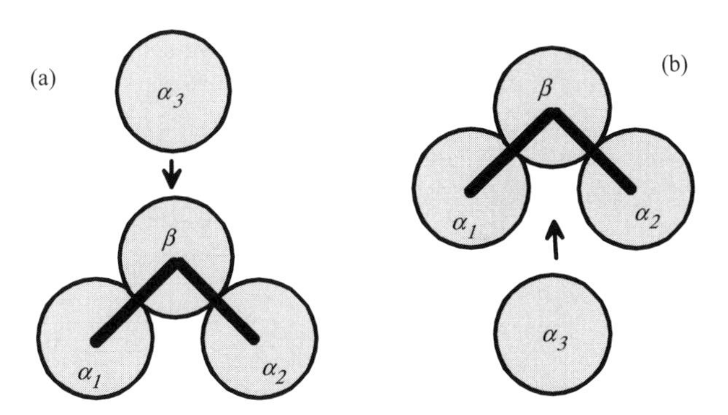


**Figure 8.** A schematic diagram illustrating the mechanism for producing compressive stress in the covalent bonds as a result of steric shielding.

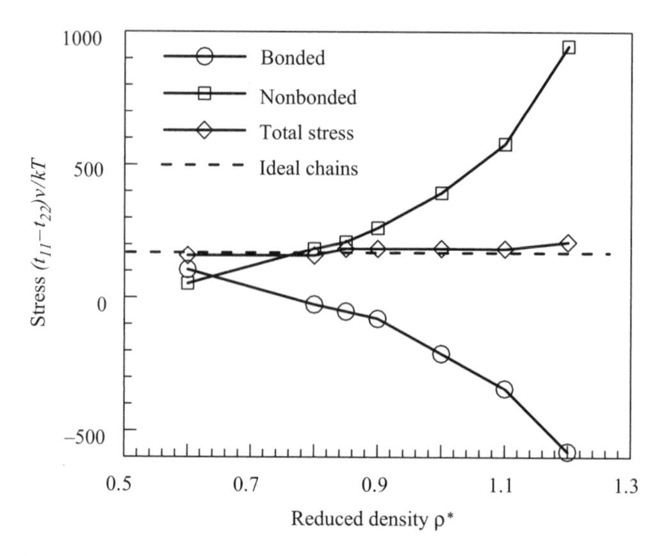


**Figure 9.** Simulated model network of $N_c = 48$ chains and $\lambda = 2$ with behavior of the contributions to the total stress $\sigma_{total}$ made by the nonbonded interactions, $\sigma_{nb}$, and by the bonded interactions, $\sigma_b.\sigma_{Ideal} = 169$, by Eq. (16).

We note from Fig. 9 that the individual bonded and nonbonded contributions to $t_{11} - t_{22}$, as they vary with $\rho^*$, are approximately equal in absolute value although opposite in sign. A possible heuristic picture for this behavior is based on Fig (8a); because of the short range and strongly repulsive character of $u_{nb}$, a nonbonded interaction has the character of a classic collision—the duration of an individual interaction is very short. As a result, the impacted atom ($\beta$ in Fig. 8a) moves little and the resultant of the impulsive compressive forces in its bonds to $\alpha_1$ and $\alpha_2$ balance that exerted by $\alpha_3$ on $\beta$. However, a collision in the configuration shown in Fig. 8b would contribute an impulsive tensile force to its bonds, but a collision between $\alpha_3$ and $\beta$ in that configuration is less likely because of steric shielding and, as verified in simulations, occurs less frequently.

A second simulation focused on the roles of the forces on a generic atom $\beta$ in a melt caused by nonbonded interactions and those caused by the forces in its connecting bonds, and on the manner in which these forces varied with the reduced density $\rho^*$ of the system: Referring to Fig. 8a, let $\boldsymbol{e}_1$ and $\boldsymbol{e}_2$ be unit vectors directed along the bonds from $\beta$ to $\alpha_1$ and $\alpha_2$, respectively. When $\beta$ is subjected to the impulsive $\boldsymbol{f}$ because of interaction through $u_{nb}$, define

$$\widehat{f}_{nb} = \frac{1}{2}(\boldsymbol{f} \cdot \boldsymbol{e}_1 + \boldsymbol{f} \cdot \boldsymbol{e}_2) \tag{34}$$

and let $f_{nb}$ be the time average of $\widehat{f}_{nb}$ averaged over all the atoms of the system. Let $f_b$ be the time average of the force in the bonds (caused by $u_b$) averaged over all the bonds in the system, and let $f = f_b + f_{nb}$. With these conventions, $f > 0$ denotes a force on $\beta$ directed toward the axis of the chain to which it belongs. Because of the bonds connected to it, a generic atom $\beta$ is constrained to move on a surface surrounding the chain. In thermal equilibrium, its average velocity satisfies $mv^2 = kT$, because it has 2 degrees of freedom on the surface. If it is, for example, constrained by a bond of length $a$ to move on the surface of a sphere of radius $a$, the constraint force $f$ satisfies $fa/kT \simeq 2$. In the ideal case, $\rho^* = 0$ and the entire constraint force $f$ must be supplied by the covalent bond, i.e., $f = f_b$. As the value of $\rho^*$ increases, $\beta$ is subjected to occasional collisions so that $f_{nb}$ increases and the bonds, $f_b$, need to supply less of the required constraint force $f$ This process continues so that at $\rho^* \cong 0.6$, $f_b$ changes from tensile to compressive. Furthermore, as $\rho^*$ increases, the average mobility of $\beta$ is decreased by its increased EV interactions with other atoms until finally at $\rho^* = 1.2$, which corresponds to a random close-packed configuration, its mobility is negligible, and the required constraint force $f = 0$. A subroutine in the equations of motion calculates the force $f$, $f_b$, and $f_{nb}$ for the values of $\rho^*$, and they are shown in Fig. 10.

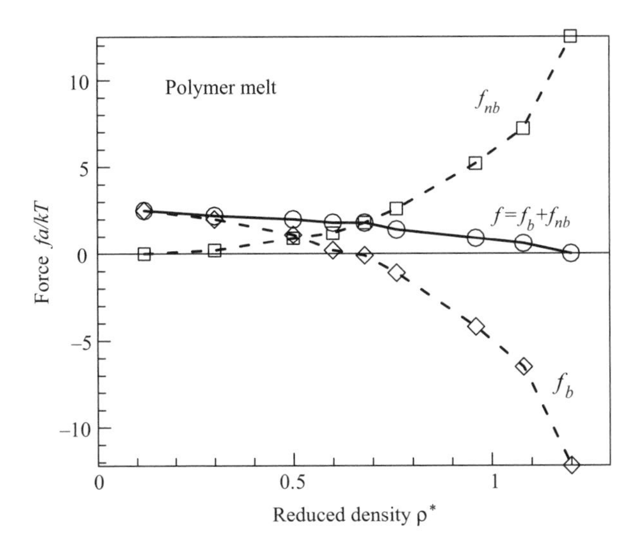


**Figure 10.** Dependence of force in bond direction in the melt as a fuction of reduced density $\rho^*$. $f_b$ in the average force in the covalent bonds, $f_{nb}$ is the average component of the noncovalent EV interaction taken in the bond direction, and $f$ is their sum.

### E. Intrinsic Monomer Stress [15]

In Sections IIB through IID, we presented important characteristics of the nonbonded interactions that contribute to the deviatoric stress. In Section II.C, it was observed that the short-range repulsive EV portion of $u_{nb}$ made the dominant contribution, whereas the attractive portion of $u_{nb}$ was of negligible importance. Furthermore, the contribution $\sigma_{ij}(\beta)$ to the deviatoric stress that is made by nonbonded interactions with atom $\beta$ depends on the steric shielding of that atom by its neighbors.

As an example, consider a chain of atoms .... $\beta - 1, \beta, \beta + 1 \ldots$ with a fixed valence angle so that the angle between bonds $\beta - 1, \beta$ and $\beta, \beta + 1$ is fixed at $\theta$. Let $e_i, i = 1, 2, 3$, be the unit base vectors of the fixed laboratory reference frame. In addition, we introduce a moving local Cartesian system with base vectors $\boldsymbol{a}_j = 1, 2, 3$, defined in Fig. 11. Then any vector $\boldsymbol{r}$ can be written in terms of components with respect to either frame as $\boldsymbol{r} = r_i\, \boldsymbol{e}_i = \bar{r}_j\, \boldsymbol{a}_j$, where we are adopting throughout the convention that superimposed bars denote components with respect to the local frame $\boldsymbol{a}_i$. The vector components with respect to the two frames are related in the usual way as

$$r_i = \bar{r}_j\, \boldsymbol{a}_j \cdot\, \boldsymbol{e}_i = \bar{r}_j a_{ji} \tag{35}$$

where $a_{ji} = \boldsymbol{a}_j \cdot \boldsymbol{e}_i$ with inverse relation,

$$\bar{r}_j = r_i a_{ij}^{-1} \tag{36}$$

where $a_{ij}^{-1} = \boldsymbol{e}_i \cdot \boldsymbol{a}_j = a_{ji}$.

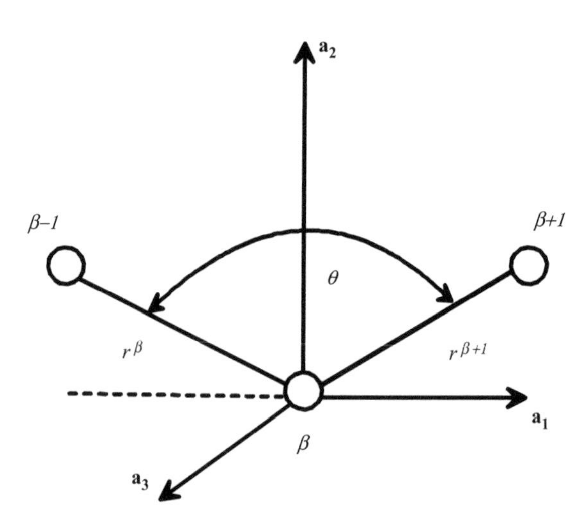


**Figure 11.** Unit orthogonal base vectors $\mathbf{a}_1$, $\mathbf{a}_2$, and $\mathbf{a}_3$ for local Cartesian coordinate system associated with chain interior atom $\beta$. Atoms $\beta - 1$, $\beta$, $\beta + 1$ lie in $\mathbf{a}_1$, $\mathbf{a}_2$ plane and $\mathbf{a}_2$ bisects bond angle $\theta$. Also shown are bond vectors $\mathbf{r}^{\beta}$ and $\mathbf{r}^{\beta+1}$.

Similarly, the atomic level stress tensor can be expressed in terms of components with respect to either frame as $\sigma(\beta) = \sigma_{ij}\boldsymbol{e}_i\boldsymbol{e}_j = \bar{\sigma}_{rs}\boldsymbol{a}_r\boldsymbol{a}_s$ with the tensor components related as

$$\sigma_{ij}(\beta) = \bar{\sigma}_{rs}(\beta)a_{ri}a_{sj}$$
$$\bar{\sigma}_{rs}(\beta) = \sigma_{ij}(\beta)a_{ir}^{-1}a_{js}^{-1}$$

where, for any chain-interior atom, $a_{ri} = a_{ri}(\beta) = \boldsymbol{a}_r(\beta) \cdot \boldsymbol{e}_i$ is the transformation matrix for the moving coordinate system attached to atom $\beta$. In the $a_r$ frame, the relative geometry of $\beta$ and its neighbors $\beta - 1$ and $\beta + 1$ is constant. Our understanding of the important role of steric shielding then leads us to expect $\bar{\sigma}_{rs}(\beta)$ to be constant.

A simulation of a collection of like chains under both melt and network conditions was performed to test this hypothesis. In the melt condition, no restrictions were placed on the chain vectors, whereas in the network condition, chain vectors were controlled and subjected to a constant volume extension $\lambda$. In both cases, $\bar{\sigma}_{rs}$ was found to be diagonal, as expected from symmetry considerations, whereas the principal values were the same, within computation accuracy, both for the strained network where they are independent of $\lambda$, and for the melt; see Fig. 12.

To obtain a stress–strain relation with this formalism, it is necessary to keep track of the effect of the deformation as it is performed on each chain $\gamma, \gamma = 1, \ldots, N_c$, in the system of $N_c$ chains with controlled chain vectors. In the initially generated systems, the chain vectors $\boldsymbol{R}(\gamma)$ have an isotropic distribution and the initial stress $\frac{\mathrm{v}t_{ij}}{kT} = \sum_{\gamma=1}^{N_c} \sigma_{ij}(\gamma)$, where $\sigma_{ij}(\gamma)$ is the stress contribution of

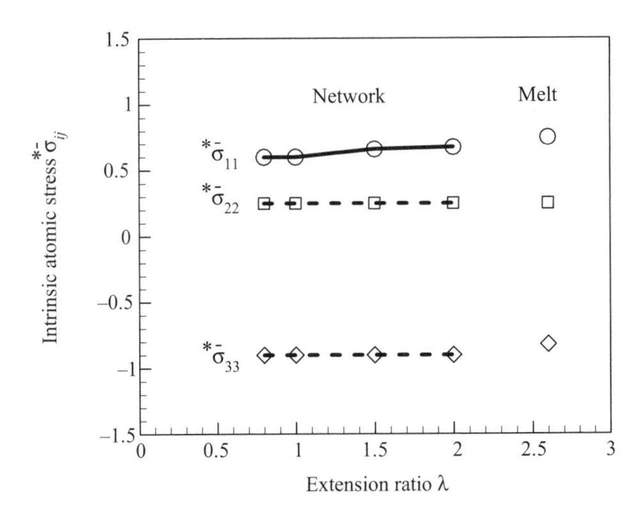


**Figure 12.** Intrinsic atomic stresses $^*\sigma_{11}$, $^*\sigma_{22}$, $^*\sigma_{33}$ as determined from molecular dynamics simulation of tetrafunctional network model in uniaxial volume deformation and for corresponding melt. (After Ref. [19].)

chain $\gamma$ and has zero deviatoric portion. The algorithm for computation of $t_{ij}(\gamma)$ is then as follows:

(1) With the intrinsic local coordinate system as in Fig. 11, the value of $\bar{\sigma}_{rs} = \bar{\sigma}_{rs}(\beta)$ averaged over all internal mers of the system is computed. Because $\bar{\sigma}_{rs}$ should be a constant for each computation, its value should be an accurate number with little fluctuation. (Except for statistical scatter, the values of $\bar{\sigma}_{rs}(\beta)$ should be equal for all internal mers.)

(2) Apply deformation $\lambda$ to all chain vectors $\boldsymbol{R}(\gamma)$ and allow the system to return to equilibrium after each change.

(3) For each chain $\gamma$, compute $\langle a_{ri}(\gamma, \lambda)\rangle$, which is the average value of the matrix $a_{ri}(\gamma, \lambda)$ for all internal mers of chain $\gamma$ at deformation $\lambda$.

(4) Compute $\sigma_{ij}(\gamma, \lambda) = \langle a_{ri}(\gamma, \lambda)\rangle \langle a_{sj}(\gamma, \lambda)\rangle \bar{\sigma}_{rs}$.

(5) $\sigma_{ij}(\lambda) = \sum_{\gamma} \sigma_{ij}(\gamma, \lambda)$.

This algorithm provides a different physical picture for the development of stress in a polymeric system.

## III. EFFECT OF AMBIENT PRESSURE

### A. Experimental program of Quested et al.

Quested et al. [16] have conducted an extensive experimental program on the stress–strain behavior of the elastomer solithane while subjected to an ambient at high pressure. Some of their experimental results are reproduced in Fig. 13. (Note that the reported stress is the deviatoric, not the total, stress as observed from the fact that the reported stress is zero for $\lambda = 1$ for the various imposed ambient pressures). For the classic ideal affine network model (all stress caused by ideal $N_c$ Gaussian chains in a volume v with no nonbonded interactions)

$$t_{11} - t_{22} = \frac{N_c kT}{\mathrm{v}}(\lambda^2 - \lambda^{-1}) \tag{37}$$

so that the only change caused by a change in ambient pressure predicted by Eq. (37) is that caused by the change in v. However, because the bulk modulus $K \cong 50\,\mathrm{kb} = 5000\,\mathrm{MN/m^2}$, the only predicted change by the ideal theory in stress at $\lambda = 1.6$ caused by a change in ambient pressure from 1 kb to 2 kb is ~2%, whereas the observed change is ~100%. Thus, these experiments provide clear evidence for the inadequacy of ideal theories that omit the contributions of nonbonded interactions. The results of simulations that do include these contributions are presented in Ref. [11].

Also observed in Fig. 13 is a change in mechanical stress–strain behavior that occurs for a change in ambient pressure between $p = 4\,\mathrm{kb}$ and $p = 5\,\mathrm{kb}$. The authors speak of this as a pressure-induced transition. Another interpretation of

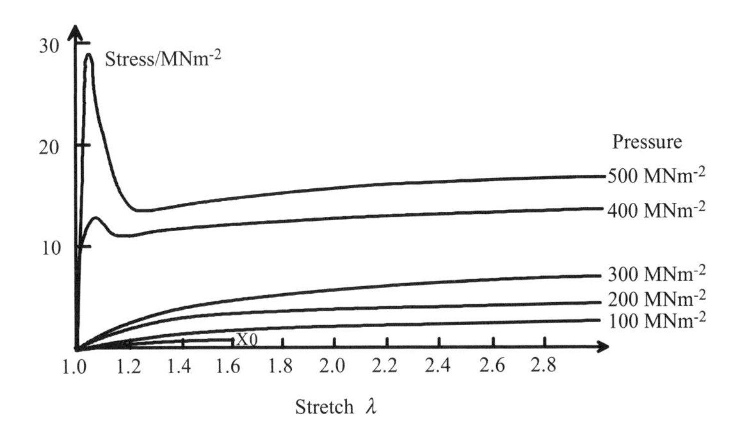


**Figure 13.** Experimental stress-stretch curves for solithane under uniaxial straining at several ambient pressures. A positive pressure corresponds to a negative mean stress. Data are re-plotted from Quested et al. Ref [16].

these results may be to postulate the existence of a relation $T_g(\eta)$ for the glass transition temperature as a function of the monomer packing fraction $\eta$ so that as the pressure increases in the interval $4\,\text{kb} < p < 5\,\text{kb}$, the value of $T_g$ increases to the temperature at which these tests are performed $(1\,\text{kb} = 100\ \text{MNm}^{-2})$.

## IV. SUMMARY

One principal difficulty in the development of a comprehensive theory of rubber elasticity is that it deals with a heterogeneous system that comprises a wide range of significant length scales, which include atomic, monomeric, mer packing fraction, chain length, junction density, network topology, and macroscopic. The monomer scale model treated here represents a small interior portion of a large system, which is suitably coupled to a large system surrounding it and is subject to the same deformation as imposed on the latter. An extensive discussion of atomic/continuum coupling in computational materials science for crystalline systems is Ref. [22]. A schematic representation of this coupling for an elastomeric system is given in Ref. [6], p. 248.

This work is based on the molecular dynamic simulation of a monomer scale model corresponding to the affine network model of rubber elasticity.[3] However, whereas the classic model has no nonbonded interactions, our model

---

[3]A treatment of the classic phantom network model is contained in Ref. [6], page 252–256. However, the statement on p. 256 that this model leads to the same stress–strain relation as the affine model is incorrect because it neglects the effect of junction fluctuations on the predicted shear modulus. See Graessley [17].

includes nonbonded interactions via the Lennard-Jones potential (Eq. 5) between all mers. The L-J parameter $\sigma_{LJ}$ is set equal to the bond length $a$ so that the mers approximate the hard spheres of diameter $a$.

1. The most important result is that the nonbonded interactions make a significant contribution to the deviatoric stress via the mechanism of anisotropic steric shielding. This mechanism is not present in network scale models in which polymer chains appear as threads without structure.
2. A significant parameter for model behavior is the reduced density $\rho^*$(Eq.(8)) of the mers. For $\rho^* > 1$, the model exhibits a glass transition at sufficiently low temperature; this transition is absent for the classic affine network model composed of noninteracting Gaussian chains.
3. Stress–strain experiments in an elastomer with controlled variation of $\rho^*$ by application of high ambient pressures have been performed by Quested et al. [16]. They demonstrate clearly the sensitivity of the deviatoric stress to $\rho^*$ and therefore testify to the importance of nonbonded interactions in its production.
4. An individual monomer and its immediate neighborhood may be likened to the interior of the basic cell of a crystal. However, whereas in a perfect crystal each cell maintains a fixed, like orientation, this orientation varies widely in a polymeric system. This difficulty can be avoided by using an appropriate intrinsic coordinate system for each mer and by determining mer orientation for the chain to which the mers in question belong. This approach leads, for example, to the concept of intrinsic monomer stress and its use to develop a stress–strain relation. It may also be used in computing the effect of a deformation on other macroscopic properties that may be associated with a single mer, e.g., optical birefringence [18].

## References

1. B. A. Boley and J. H. Weiner in Theory of Thermal Stresses, Dover, UK, 1997.
2. L. R. G. Treloar, The Physics of Rubber Elasticity, 3rd ed., Clarendon Press, Oxford, 1975.
3. G. Ciccotti, D. Frenkel, and J. R. McDonald, *Simulation of Liquids and Solids*, North-Holland 1987.
4. J. Gao and J. H. Weiner, *Macromolecules* **20**, 2520 (1987); **20**, 2525 (1987); **21**, 773 (1988); **22**, 979(1989).
5. R. J. Swenson, *Am. J. Phy.* **51**, 940 (1983).
6. J. H. Weiner, *Statistical Mechanics of Elasticity*, 2nd ed., Dover, UK, 2002.
7. A. F. Bower and J. H. Weiner, *J. Chem. Phys.* **118**, 11297 (2003).
8. D. W. van Krevelen, *Properties of Polymers*, 3rd ed., Elsevier, New York, 1972, pp 44–54. (The data used here, Table 4.3 on p. 44, are titled "Molar volumes of rubbery amorphous polymers at 25° C." They need not necessarily represent crosslinked elastomers. We are making an

assumption that these data are also relevant in our study of rubber elasticity; see Fig. 9.5 and discussion on p. 159 of this reference.)

9. P. Chaikin in *Soft and Fragile Matter*, M. E. Cates and M. R. Evans, eds., Institute of Physics Publishing, Philadelphia, PA, 2000, p. 320. (See also D. Chandler, J. D. Weeks, and H. C. Andersen, *Science* **220**, 787 (1983), for a discussion of the van der Waals picture of liquids and solids. See, in particular, the graphic description under the heading, "The Basic Idea" on p. 787.)
10. H. M. James and E. Guth, *J. Chem. Phys.* **11**, 455–481 (1943).
11. A. F. Bower and J. H. Weiner, *J. Chem. Phys.* **120**, 11948 (2004).
12. J. Gao and J. H. Weiner, *Macromolecules* **24**, 5179 (1991).
13. D. A. McQuarrie, *Statistical Mechanics*, Harper and Row, New York, 1976, p. 266.
14. S. P. Hansen and I. R. McDonald, Theory of Simple Liquids, 2nd. ed. Academic Press, New York, 1986.
15. J. Gao and J. H. Weiner, *J. Chem. Phys.* **90**, 6749 (1989).
16. D. L. Quested, K. D. Pae, J. L. Sheinbein, and B. A. Newman, *J. Appl. Phys.*, **52**, 5977 (1981).
17. W. W. Graessley, *Polymeric Liquids and Networks: Structure and Properties*, Garland Science: Taylor and Francis Group, New York, 2004, pp. 447–451.
18. J. Gao and J. H. Weiner *Macromolecules*, **27**, 1201 (1994).
19. J. H. Weiner and J. Gao, in Computer Simulation of Polymers, R.J. Roe ed., Prentice Hall, New York, 1991, pp. 255–261.
20. A. F. Bower and J. H. Weiner, *J. Chem. Phys.* **125**, 096101 (2006).
21. J. Gao and J. H. Weiner, *Macromolecules* **22**, 979 (1989).
22. W. A. Curtin and R. E. Miller, *Modelling Simul. Mater. Sci. Eng.* **11**, R33-R68.

# MULTIPLE AMORPHOUS–AMORPHOUS TRANSITIONS

THOMAS LOERTING

*Institute of Physical Chemistry, University of Innsbruck, Innrain 52a, A-6020 Innsbruck, Austria*

VADIM V. BRAZHKIN

*Institute for High Pressure Physics, Troitsk, Moscow Region, 142190 Russia*

TETSUYA MORISHITA

*Research Institute for Computational Sciences (RICS), National Institute of Advanced Industrial Science and Technology (AIST), Central 2, 1-1-1 Umezono, Tsukuba, Ibaraki, 305-8568 Japan*

## CONTENTS



*Advances in Chemical Physics, Volume 143*, edited by Stuart A. Rice

# I. INTRODUCTION

The amorphous solid state of matter has traditionally received much less attention than the crystalline solid state [1–4]. By contrast to crystalline solids, which are ordered and can be defined using a periodically repeated unit cell, amorphous solids are disordered on long-range scales. Amorphous solids show a short-range order, e.g., tetrahedrality, which is similar to the short-range order found in crystalline solids. However, long-range order in amorphous solids does not exist because order typically disappears at distances of 20–50 Å. Because of their inherent disorder, amorphous solids are metastable with respect to the well-ordered crystals. The difference in Gibbs free energies is often called "excess free energy." Many methods of producing amorphous solids are available using the gas, the liquid, and the solid as starting materials. All of them share the principle that excess free energy has to be provided, e.g., by mechanical, thermal, or chemical treatment, which is taken up by the material [5]. Amorphous solids are often divided in two subgroups—glasses and nonglassy amorphous solids. One of the most traditional routes of producing glasses is by cooling the liquid below the melting temperature without crystallizing it [6, 7]. Thus, glasses are synonymously also called vitrified liquids. After reheating a glassy solid, it turns into a supercooled liquid above the glass→liquid transition temperature $T_g$. In the transition region from the glassy solid to the supercooled liquid near $T_g$ the microscopic structure does not change—rather the relaxation dynamics change [8]. Whereas glasses are nonequilibrium states, which slowly relax toward the metastable equilibrium on the experimental time scale, the metastable equilibrium is reached within the experimental time scale in case of supercooled liquids. By contrast, a nonglassy amorphous solid does not turn into a supercooled liquid during heating. Instead, either it remains in the amorphous solid state or it crystallizes. As an explanation why these amorphous solids behave unlike glasses, often the concept of "nanocrystalline" material is invoked [9–11], which basically implies that the material is made of a huge number of very small crystal grains, each of which contains on the order of a few hundred molecules. In this view, the sharp Bragg peaks in scattering experiments characteristic of crystalline material are not observed because they are massively broadened because of the small crystal sizes. For example, the X-ray pattern cannot be distinguished then from true glassy material, and thus, the "nanocrystalline" material looks "X-ray amorphous." Although both "nanocrystalline" and "glassy" solids do not show long-range order, they can be distinguished in terms of how the short-range order disappears at intermediate ranges [12]. In the glassy state, the order disappears more or less continuously by variation of interbond angles and (to a lesser extent) interatomic distances. In the "nanocrystalline" state, the order is lost more or less discontinuously at the grain boundaries—almost no variation of interbond angles and interatomic distances is observed inside the nanocrystals, but strong variation exists at the intergrain boundaries (although the size of the nanograin and the intergrain boundary may be compatible to each other). Instead of grain boundaries, defective crystals distorted by linear defects (disclinations, dispirations, and displanations) have been considered [13]. So, in amorphs one speaks about three distance ranges as follows: (1) the short-range order, which typically reaches the direct neighbors of a central atom or molecule, i.e., the first coordination sphere (sometimes the second coordination sphere); (2) the intermediate range order, which typically encompasses the second to the fifth through seventh coordination spheres, where the loss of order occurs [14–20]; and (3) the long range beyond the fifth through seventh coordination spheres, where there is no order.

During variation of pressure and/or temperature, the amorphous solids may experience different processes, which can be superimposed as follows: (1) elastic reversible behavior, that is, the connectivity in the network remains the same—just isothermal compression and/or thermal expansion take place; (2) inelastic behavior associated with the change of intermediate-range order; (3) inelastic behavior associated with a change of short-range order including coordination changes in the first through second spheres; and (4) irreversible structural relaxation associated with the equilibration process toward the metastable equilibrium. Whereas (1) is found for any material and (4) is related to the kinetics of relaxation, (2) and especially (3) represent structural transitions governed by the laws of thermodynamics analogous to phase transitions for crystals. In this sense, structural transitions in amorphs have been called "phase" transitions, even though amorph (nonequilibrium nature) and phase (equilibrium nature) are in direct contradiction. Glasses and amorphous solids are nonergodic, non-equilibrium states, and a strict physical meaning of the "phase" can be applied to them only conditionally ("metastable phase") [21]. The use of terminology reserved for true equilibrium transitions (phase, latent heat, first-order nature, nucleation, etc.) seems to be justified in view of the experimental finding that the metastable amorphous states can be stabilized for very long times, if not infinitely, under suitable conditions [i.e., process (4) can be suppressed effectively]. That is, the term "phase" can be applied conditionally for "metastable phase." In this sense, the amorphs can be viewed to be in a metastable quasiequilibrium rather than in nonequilibrium. Such a metastable quasiequilibrium allows for the definition of a coexistence line, where Gibbs free energies of two amorphous states are identical, and the possibility of "first-order" transitions that involve jump-like changes in entropy, volume, and enthalpy arises. Also the possibility of a critical point located at the end of the coexistence line arises. Another complicating circumstance with respect to amorphous solids and glasses is that the transformations between different "phases" during the experimental times occur, as a rule, under conditions that are far from equilibrium; these transformations are determined by kinetic parameters, much like low-temperature phase transitions with a large hysteresis found in crystals [22].

Structural phase transitions in crystals under changes of the P,T-parameters are a well-studied phenomenon from both the experimental and theoretical viewpoints. The concept of "polymorphism," i.e., the occurrence of more than one crystal structure for a single component, is a well-known and traditional concept. For many components, the phase diagram of stable polymorphs has been constructed and shows lines in the P,T diagram, where two polymorphs can coexist in thermodynamic equilibrium. Crossing these lines then causes a sharp structural transition, i.e., a change of short-range order in the unit cell [very much like mechanism (3)] [23]. However, the concept of "poly-amorphism," i.e., the occurrence of more than one amorph, and structural transitions between these amorphs is relatively new and not well understood. Disordered systems possess additional, more essential characteristics that differentiate them from crystals, namely, the inhomogeneity of their structure and the dispersion of their properties on the scales smaller than the correlation length of a medium order in disordered systems. For this reason, the disordered systems follow two scenarios of phase transformations: Either they can experience a first-order phase transition, provided the minimum size of a nucleus of an emerging phase is larger than the correlation length of a disordered state, or, by contrast, they undergo smeared changes (at all temperatures) [24]. For the smooth transitions, there is a series of intermediate states, each being in compliance with the condition of minimum of the Gibbs free energy. This particular scenario cannot be realized in crystals, where just two definite phases exist whose macroscopic mixture may be thermodynamically equilibrated in the transition point only. Thus, the experimental observation of sharp changes of the structure and properties in the disordered systems, as opposed to crystals, is a necessary but insufficient evidence for a first-order transition, for there always remains the possibility of this transition occurring in a fairly narrow but finite P,T-interval. As a result, the principal evidence for the first-order transition in the disordered media is the observation of a macroscopic mixture of phases during transition [22].

The concept of a *single* structural transition in amorphous material, i.e., an amorphous–amorphous transition, was coined in 1985 by Mishima et al. [25] on the example of water. Today, in many respects the nature of pressure- and/or temperature-induced transformations in glasses and amorphous solids remains unclear. In most cases, the pressure treatment of glasses and amorphous solids results in residual densification. In the process, the densified glasses and amorphous solids do not undergo any significant change in a short-range order (coordination number Z) as a rule; it is only amorphous network topology (ring statistics, etc.) that varies. Nevertheless, there are some examples of coordination transformations in amorphous solids, glasses, and liquids, which can take place in a sharp manner or smeared over a wide pressure and temperature range, e.g., amorphous $H_2O$, $SiO_2$, $GeO_2$, $B_2O_3$, Si, Ge (for references, see Sections II through IV), $A_3B_5$-compounds, Se, S, P, $AlCl_3$, $ZnCl_2$, ZnSe, amorphous irradiated zircon [26], some metallic glasses (e.g., $Ce_{45}Al_{55}$) [27], and in aluminates [28]. These examples, where a coordination transformation in the amorphous solid can be observed, are the candidates for possible *multiple* amorphous–amorphous transitions, which are the focus of the current review. Among these candidates, some evidence has been accreted for a possible multiple nature of amorphous–amorphous transitions. Therefore, we discuss covalent oxide glasses such as $SiO_2$, $GeO_2$, and $B_2O_3$ in Section II, $H_2O$ in Section III, and the semiconductors Si and Ge in Section IV. Interestingly, a good correlation exists between the few glasses that show coordination transformations and the few liquids known to show anomalous behavior, such as an apparently diverging heat capacity and compressibility during supercooling, the negative melting slope in the P-T phase diagram, or a density maximum. Although the possibility of structural transformations under compression in some glasses is beyond doubt, the question whether the phase diagram of the glass (and melt) is a shifted reflection of the pressure–temperature phase diagram of the crystal remains to be answered. Multiple-phase transitions under pressure for one substance in the crystalline state are possible often; whether multiple structural transformations can similarly be observed in a respective glass and melt is a matter of debate. Whereas there is barely any experimental information on true multiple liquid–liquid transitions (except for the case of AsS [29]), simulations have suggested the possibility of multiple first-order liquid–liquid transitions [30–36].

Although we focus here on the *status quo* concerning the debated aspect of multiple amorphous–amorphous structural transitions, we want to emphasize that there are numerous reviews on the topic of polyamorphism and single amorphous–amorphous transitions worth reading [37–48].

## II. COVALENT OXIDE GLASSES

### A. Silica

Of greatest interest is the study of pressure-induced transformations in such an archetypical glass as a-$SiO_2$. Besides the fundamental importance of research on transformations in disordered media, a deeper insight into the transitions that occur in a-$SiO_2$ and other amorphous and liquid silicates is of much importance for the physics of the Earth and planetary interiors. Thus, we pay the most attention to the a-$SiO_2$ study, whereas other oxide glasses such as $GeO_2$ and $B_2O_3$ will be considered more concisely.

Silica ($SiO_2$) is the substance whose properties and phase-transition behavior are vital for understanding the puzzling processes and phenomena existing in the Earth's upper mantle and in Earth-like planets [49, 50]. Silica, being one of the most abundant components of the Earth, "plays a major role in the deep interior, both as a product of chemical reactions and as an important secondary phase" [51]. The equilibrium pressure-temperature (P-T) phase diagram of silica is relatively well understood, and it has been extensively reviewed elsewhere [51] (Fig. 1). Under compression, the $SiO_2$ crystal undergoes a set of transitions from α-quartz (the silicon atom coordination number relative to oxygen atom number, $Z = 4$) $\rightarrow$ coesite ($Z = 4$) ($P \sim 3$ GPa) $\rightarrow$ stishovite (rutile structure) ($Z = 6$) ($P \sim 9$ GPa) $\rightarrow$ $CaCl_2$-structure type ($Z = 6$) ($P \sim 70$ GPa) $\rightarrow$ α-$PbO_2$-structure type ($Z = 6$) ($P \sim 100$ GPa) – pyrite-structure type ($Z = 8$) ($P \sim 260$ GPa)[51–63].

The behavior of the $SiO_2$ glass under pressure has been intensely studied as well [39, 53, 64–83]. It was established long ago that the high-pressure, high-temperature treatment causes a significant residual densification of silica glass

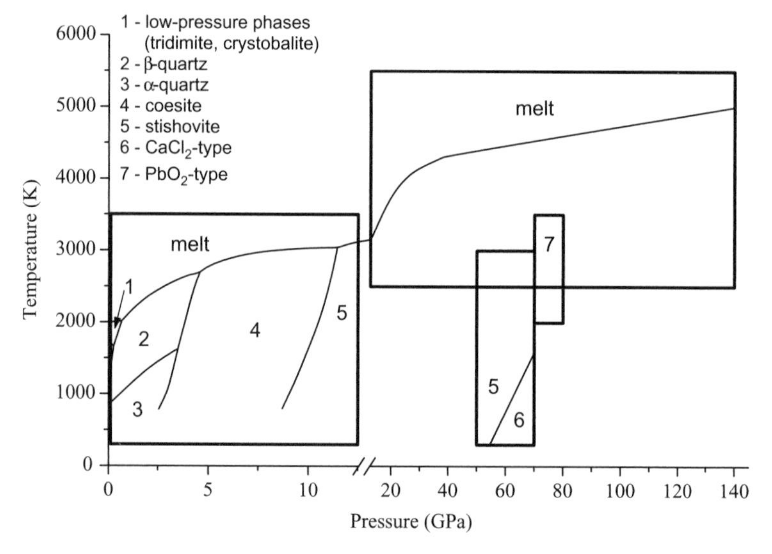


**Figure 1.** Schematic high P-T phase diagram of $SiO_2$ determined from static compression experiments. Rectangles are areas already studied. The bottom left rectangle (with 1–5 phases) is part of a diagram by both in situ and quenching techniques in a large volume press apparatus [51–54]. The upper right rectangle was studied by shock-wave and laser heating in diamond anvil study of the melting curve [55–57]. The bottom right middle rectangle with 5/6 is the study (laser heating with diamond anvils) of stishovite to $CaCl_2$ transition [58] with a line of 5/6 phase transition [59]. The bottom right rectangle with 7 is an area investigated by which includes Dubrovinsky et al. [60], laser heating with diamond anvils.

[14, 66–68, 70, 71, 76, 84–87]. The maximum densification of 18–20% is achieved after a pressure of 16–20 GPa—room temperature treatment or after 5–8 GPa—800–1000 K treatment. The densified glasses have predominantly tetrahedral coordination of silica atoms, Z $\sim$ 4–4.5 [14, 71, 88]. Elastic moduli and optical characteristics of densified glasses are distinctly different from those of pristine silica glasses.

The extensive studies on the structure [72, 89] and Raman and Brillouin spectra [68–70, 73], as well as computer simulation results [77–82] have revealed that in the 8–50 GPa pressure range and at room temperatures, the silica glass is subject to a broad transformation accompanied by a change in the short-range order structure and an increase in the coordination number from 4 to 6. It should be noted that during coordination transformation at intermediate pressures, many silicon atoms have a fivefold coordination. The main part of the transformation takes place in a narrower pressure range of 10–40 GPa.

The earlier data did not shed light on how closely this coordination transformation is related to residual densification of the glass. Another unanswered question was whether glass densification after 20 GPa pressure at room-temperatures treatment and the one at 5–8 GPa and high-temperatures treatment are of a similar nature.

Recent years have witnessed a significant advance in the understanding of the behavior of glassy silica under compression. By using a strain gauge technique, direct in situ measurements of the relative volume of the glass in a wide range of pressures (0–10 GPa) and temperatures (30–730 K) have been conducted (see Fig. 2) [90]. It has been found that at increased temperatures and pressures of 4–8 GPa, the a-$SiO_2$ glass undergoes a transformation accompanied by moderate volume changes of 2–7%, depending on the pressure. The computer simulation data [91, 92] perfectly fit those obtained by volumetric measurements (see Fig. 3). Here, we note that the data in Ref. [85] on the existence of a sharp volumetric anomaly ($\sim$20%) at 3.6 GPa and 850 K are not corroborated by the results of other independent measurements. High residual densification ($\sim$20%) in the glasses after decompression is caused by a considerably lower (by almost two times) compressibility of a densified glass [90]. Thus, the position in the T,P-plane of the zone of the transformation responsible for densification was determined (Fig. 4). A recently conducted in situ structural study on the $SiO_2$ glass under high-pressure, high-temperature conditions [17] (see Fig. 5) as well has revealed the existence of this transformation region, which is in perfect agreement with the volumetric data (see Fig. 4). The computer simulation data [91, 92] and in situ X-diffraction study [17] have allowed us to clarify the nature of this transformation. The coordination number during this transformation changes insignificantly, whereas the topology of tetrahedron connectivity substantially changes toward more efficient packing of the tetrahedra. Thus, the changes during this transformation primarily affect the

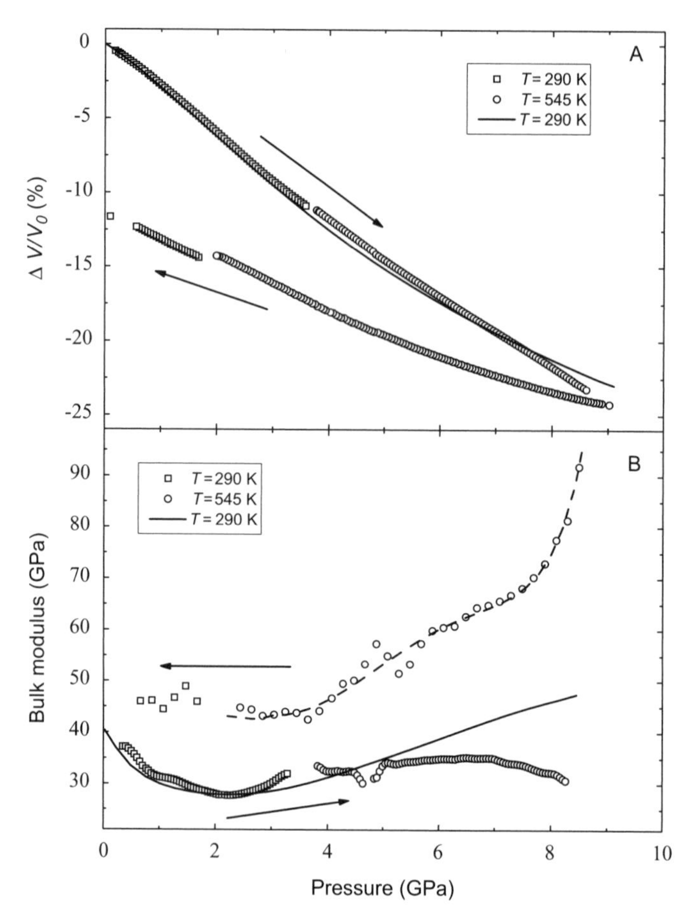


**Figure 2.** Relative change of a-$SiO_2$ volume in the compression and decompression cycles at 545 K (a) and the corresponding bulk modulus obtained by the numerical differentiation of volume curves (b). The data corresponding to the a-$SiO_2$ compression at room temperature are given for the comparison (from Ref. [90]).

intermediate-range order rather than the short-range one. In terms of the "rigidity percolation" [93], the transformation of the amorphous network with floppy modes toward the rigid network occurs [92]. The analysis of structural data has revealed that the number of small rings in the network increases during transformation; this change in the ring statistics leads to the decrease in the number and size of interstitial voids in the network. Apart from a change in the intermediate-range order, the first transformation also involves a change in the coordination of some part of silicon atoms in the fivefold and sixfold coordination states.

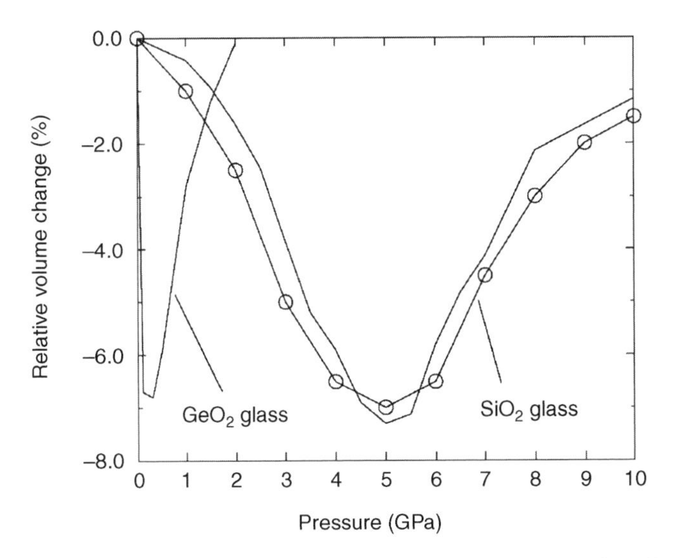


**Figure 3.** Temperature-induced densification in $SiO_2$ and $GeO_2$ glasses from Ref. [91]. A circled line is the experimental results; solid lines are the results from MD simulation.

Thus, the first transformation in a-$SiO_2$ happens with a small change in the coordination number and is irreversible at room temperatures, which gives rise to residual densification (see Fig. 4). The second transformation, accompanied by the further increase of the first coordination number up to $Z = 6$ and occurring already within the rigid network, takes place at higher pressures; it is almost independent of the temperature (see Fig. 4) and is nearly fully reversible. At room temperature, these two transformations significantly pressure overlap, whereas at higher temperatures, they can be conditionally separated.

The pressure range for the second transformation is still debatable. Note that the recent study on the bonding changes in the $SiO_2$ glass conducted by using an inelastic X-ray scattering technique [94] gives an unreliable estimate of the degree of coordination transformation. According to Ref. [94], the transformation in the $SiO_2$ glass to the sixfold coordination state of silicon atoms is completed at $P \sim 20$ GPa. At the same time, the direct structural study of the $SiO_2$ glass indicates that this coordination transformation is not completed even at $P \sim 40$ GPa [72, 89]. At $P \sim 18$ GPa, the coordination changes only begin [17]. A very recent inelastic X-ray Si L-edge scattering study has revealed that there is no domination of six coordinated Si atoms even at 74 GPa [95]. Fukui et al. [95] suggested that a significant fraction of five coordinated Si atoms exist up to the highest pressures. Thus, the assessments of the degree of coordination transformation in glasses made from inelastic X-ray scattering data (e.g., Refs. [94, 96, 97]) should be treated with caution.

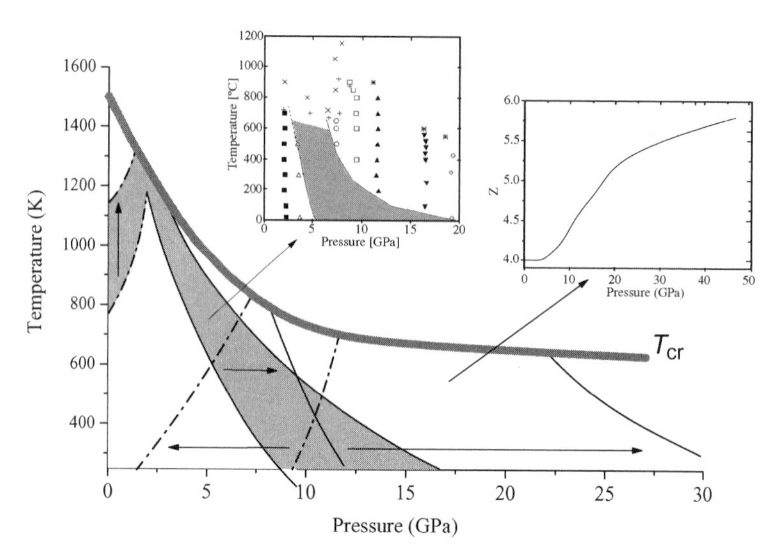


**Figure 4.** Phase diagram for the transformations of glassy $SiO_2$ according to the volumetric data from the Ref. [90]. These data are evidence for the two transitions: between the ordinary glass and densified rigid network state of glass with a small increase of coordination (the gray regions correspond to the direct and reverse transitions) and between densified state of glass and six-fold coordinated phase (nonhatched regions). The regions of direct transitions are bounded by the solid lines, and the regions of reverse transitions are bounded by dot-and-dash lines. The arrows show the directions of the corresponding transitions. The insets show the region of the first transition according to the structural study of Ref. [17] (giving with the authors' permission) and change of Si-atoms coordination during the second transition according to the computer simulation study from Ref. [80].

We note that the first transformation results in the formation of a rigid unstressed state of the amorphous network [91]. As the coordination number increases, the amorphous network becomes overstressed and overconstrained. In the case of a high atomic coordination, bond length and angle fluctuations, which are intrinsic for amorphous network, cost very high elastic energy. This leads first to higher transformation pressures as compared with crystalline prototypes and, second to the impossibility of retaining a high coordination number after decompression, which is another fact that strongly differentiates the glasses from their crystalline counterparts. Finally, these high elastic stresses in overconstrained amorphous networks cause a higher tendency to nano-crystallization of such glasses.

This is, even to a greater extent, true of oxide glasses with the coordination number $Z > 6$. It is obvious that with another increase in pressure to $P \sim 1$ Mbar (the stability region of an $\alpha$-$PbO_2$ structure type), the coordination number in glassy silica should slightly rise [98] and approach eight at $P \sim 2$–3 Mbar (the stability region of a pyrite-like structure of crystalline silica). This state of the

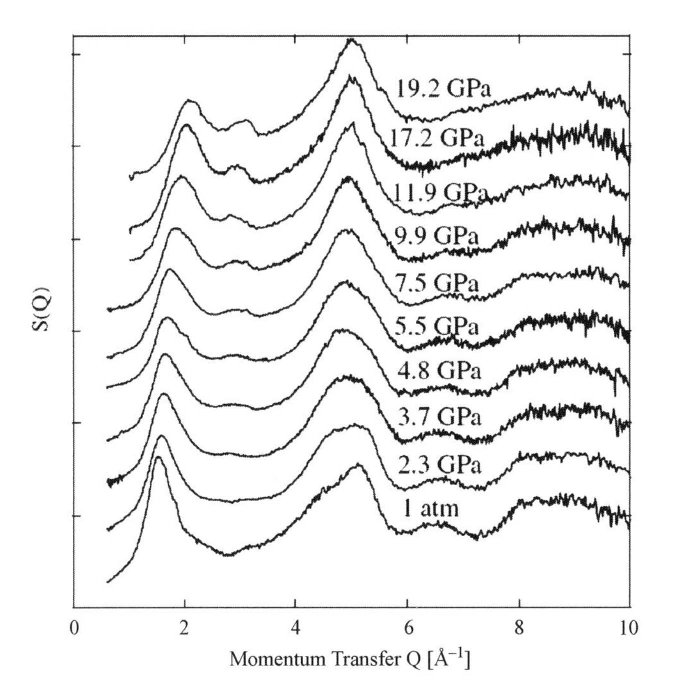


**Figure 5.** The pressure dependence of the structure factor for $SiO_2$ glass at room temperature (from Ref. [17] with the authors' permission)

glass should be overstressed even more and should have a strong tendency to crystallization at nanoscales.

The presence of several structural phase transformations in the glassy state permits suggesting the respective structural changes in the melt. Thus, for example, in liquid $SiO_2$, the smooth transition with a change in the intermediate-range structure at $P \sim 3$–5 GPa is manifested in the degree of densification of glasses obtained by quenching from melt under pressure [99], whereas the other transition is associated with a considerable rearrangement of the atomic short-range order and accounts for the anomalies of the properties of the $SiO_2$ melt [100–105].

## B. Other Oxide Glasses

The $SiO_2$ glass seems to be the most well-known glass but is not the only one to reveal the complicated behavior of its structural characteristics under compression. The $GeO_2$ substance is analogous to $SiO_2$. The P,T diagram of $GeO_2$ is similar to that of $SiO_2$ (in fact, even more simple as it lacks many four coordinated phases, which includes a coesite-like one). A rutile-like phase of $GeO_2$ ($Z = 6$) transits to a $CaCl_2$ structure type ($Z = 6$) ($P \sim 30$ GPa), then to an

$\alpha$-$PbO_2$-structure type (Z = 6) (P ~ 50 GPa), and finally to a pyrite-like structure (Z = 8) (P ~ 85 GPa) [63, 106, 107]. As with a-$SiO_2$, the P,T-treatment of the $GeO_2$ glass brings about residual densification of ~14–16% without a significant change of short-range order coordination [16, 108].

The coordination transformation in glassy $GeO_2$ under compression with an increase in Z from 4 to 6 takes place in the 5–20-GPa pressure range, which is attested by both experimental data of the extended X-ray absorption fine structure (EXAFS) [109], volumetric [76, 110], and X-ray diffraction [111, 112] measurements, as well as computer simulation result [113, 114]. The coordination number in $\alpha$-$GeO_2$ rises nonmonotonically with pressure: At P ~9–10 GPa, the rate of rise of the Z value slows down [112]. According to Guthrie et al. an intermediate fivefold coordinated state of $\alpha$-$GeO_2$, exsits; however, it is more likely that this intermediate state includes Ge atoms of different coordinations, of 4, 5, and 6 [113, 114]. Recent structural studies on $\alpha$-$GeO_2$ performed simultaneously at high pressures and temperatures obtained at isobaric heating [115] and isothermal pressure runs [116] have revealed a complicated picture of structural changes. Similar to a-$SiO_2$, the a-$GeO_2$ substance in the region of moderate pressures (1–4 GPa) and high temperatures (400–600 K) experiences the intermediate-range order changes accompanied only by an insignificant increase in the coordination number [115, 116]. The computer simulation study of the $GeO_2$ glass has also revealed a transition to the rigid network state at heating at P ~ 1 GPa [91] (see Fig. 3).

Thus, for a-$GeO_2$ at high temperatures, we may conditionally consider the existence of two overlapping structural transformations under compression. It is obvious that at additional increases in pressure, a-$GeO_2$ is bound to undergo, similarly to a-$SiO_2$, transformations accompanied by the rise in the coordination number to eight [98]. Glassy $GeO_2$ oxides with Z ~ 6 and Z ~ 8 are overstressed and cannot be quenched to normal pressure conditions.

The presence of several structural phase transformations in glassy $GeO_2$ suggests the respective structural changes in the melt, which is already partially confirmed by both experimental study [117] and computer simulation study [118].

$B_2O_3$ represents an archetypical oxide glass alongside such glasses as $SiO_2$ and $GeO_2$ [119], with the important difference that at the ambient conditions, the structural units are planar $BO_3$ triangles as opposed to tetrahedra. As was found long ago [120, 121], pressure-treated $B_2O_3$ glasses show residual densification $\Delta\rho/\rho \sim 5$–10%, which depend on the pressure-temperature conditions of treatment. The ex situ studies of densified glasses point to the breakup of the boroxol rings in the glass structure and to the buckling of the "ribbons" formed by the $BO_3$ triangles, without any significant coordination change of the boron and oxygen atoms [122, 123]. The in situ investigations of $B_2O_3$ glass under pressure have been performed using Raman and Brillouin spectroscopies [124–126] and through inelastic X-ray scattering spectroscopy

[97]. In addition, there have been attempts to examine the $B_2O_3$ glass under pressure by molecular dynamics computer simulation, using empirical interatomic potentials [127, 128]. According to the data of Nicholas et al. [125], $B_2O_3$ glass under compression experiences a transformation in the pressure range $P \sim 6$–15 GPa; according to Ref. [97] $B_2O_3$ glass, under compression, features a considerable change in the bonding type in the 6–20-GPa pressure range. These changes were attributed to the modification of the short-range order in glass, with the coordination of boron atoms increasing from 3 to 4 in a similar way to crystalline phases.

We have recently reported the results of the in situ diffraction experiments and the in situ volumetric measurements, and we have complemented them with the data from ab initio calculations [129].

The behavior of $B_2O_3$ glass under pressure is in a certain sense similar to the behavior of other archetypical oxide glasses, a-$SiO_2$ and $GeO_2$. Under pressure, $B_2O_3$ glass equally features two overlapping diffuse transformations (see Fig. 6). The first transformation is irreversible at room temperature without the change in the first coordination number and is accompanied by the change of

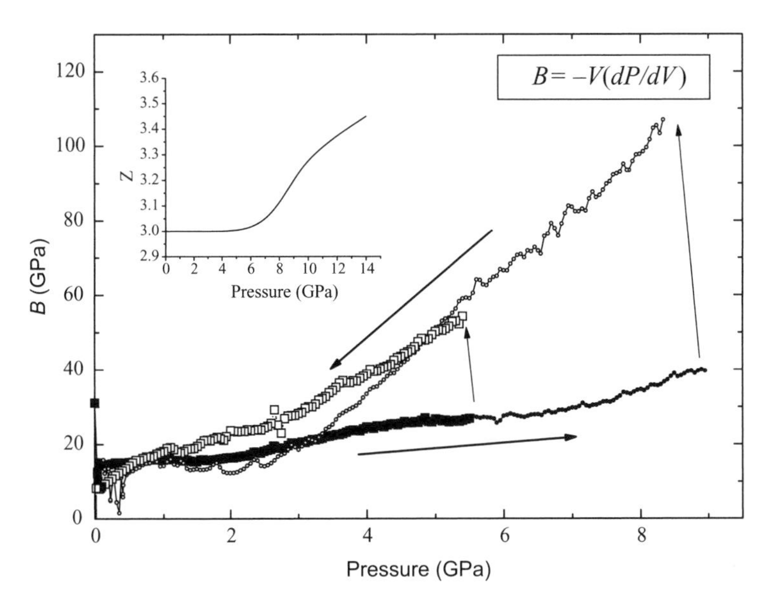


**Figure 6.** Pressure dependencies of the bulk modulus obtained by the direct numerical differentiation of the in situ volumetric measurements of the glassy $B_2O_3$ under pressure ("relaxed" modulus) in the two different runs of compression (solid symbols) and decompression (open symbols). The significant jumps of the effective bulk modulus between the final of compression and onset of decompression for both runs correspond to the jumps between "relaxed" and almost "unrelaxed" values. The inset shows pressure dependences of the first coordination number for B from the recent X-ray diffraction data. Both data are from Ref. [129].

intermediate range order; the second transformation is reversible, and it occurs at higher pressures and is accompanied by the change in the coordination of the B atoms from 3 to 4. In $B_2O_3$ glass, the increased coordination number of cation atoms is not retained at room temperature after decompression either. Similar to silica, the behavior of the glass greatly differs from that of its crystalline counterpart: The $B_2O_3$ II phase can be retained at normal conditions and has high temperature stability. Ab initio simulations predict one more transition to the glassy state with sixfold coordinated B in the megabar pressure region [129].

Thus, in all discussed oxides, the structural transformations under pressure can be described as a multiple set of transformations. First, changes occur in the intermediate-range order with a slight increase in the coordination number and a transition to the rigid network state; then, a coordination transformation with the increase of the first coordination number follows. According to the computer simulation data, additional buildup of pressure to the megabar range causes a subsequent complementary increase in the coordination number.

It is evident that similar sets of multiple-phase transitions are bound to be observed in the respective melts. Such structural studies on the oxide melts have not yet been carried out, which is because of greater experiment complexity. Nevertheless, the chalcogenide melts, which experience structural transformations at more moderate temperatures and pressures, give examples of multiple phase transitions. Thus, the AsS [29] and CdTe [130] melts feature at least two pressure-induced transitions accompanied by a change in the short-range order structure and properties of the liquids.

## III. WATER

### A. Preparation of Amorphous Ices

Metastable amorphous solids can in general be prepared from stable phases by bringing in excess free energy [5]. In the case of water, amorphous solids have been prepared from stable phases in all three aggregate states: from the gas, the liquid, and the crystalline solid [131].

#### *1. Preparation from the Gas Phase: Amorphous Solid Water (ASW)*

The formation of an amorphous solid was first reported in 1935 [132, 133]. These authors used the route of depositing warm water vapor on a cold substrate, which freezes in excess free energy by the rapid change in temperature. At substrate temperatures above $\sim$160 K, the deposit was found to be crystalline ice I, whereas below this temperature, an amorphous solid was obtained. These deposits are referred to as ASW, which is a microporous material that can adsorb gases [134, 135]. In fact, ASW also condenses on interstellar dust particles and is likely the most abundant form of solid water in the universe. Therefore, studies on ASW bear an astrophysical relevance [134, 136]. The microporosity can be reduced greatly by sintering the sample to no more than 120 K.

The density of the ASW films depends highly on the experimental conditions, e.g., on the angle of incidence of the molecular beam. For a deposition temperature of 22 K and at a normal angle, a maximum density of 0.94 $g/cm^3$ is observed, whereas a density of 0.16 $g/cm^3$ is observed close to glancing incidence, which implies a porosity of 80% [137]. ASW anneals or relaxes during heating in vacuo to a structural state approaching that of hyperquenched glassy water (HGW) [138].

Using different deposition rates, even a highly compacted form of amorphous solid water of density $>1$ $g/cm^3$ could be obtained at deposition temperature $T < 30$ K, which transforms gradually in the temperature range 38–68 K to the lower density form of density 0.94 $g/cm^3$ [139, 140]. This transition was proposed to be at the origin of crack-formation processes in comets [141]. We note, however, that the formation of this high-density amorph at very low temperatures has been doubted [142, 143]. Only photolysis at 20 K induces a transition to a high-density amorph [143].

### 2. *Preparation from the Liquid Phase: HGW*

Instead of water vapor, also the liquid can be turned into an amorphous solid by cooling. Quenching is a standard method of glass formation for many substances, both organic and inorganic in nature [3, 7]. In fact, it is believed that all liquids can, in principle, be vitrified by cooling. Some liquids can be vitrified easily even by slow cooling ("good glass formers"), whereas others can be vitrified only with difficulties by very rapid cooling ("bad glass formers"). Water is a particularly bad glass former, and a cooling rate of the order of $10^6$–$10^7$ K/s is necessary for avoiding crystallization to ice I. Achieving such high cooling rates had required the invention of new techniques, which are called "hyperquenching" or "splat cooling" techniques. Mayer and Brueggeller were the first to succeed by projecting a thin jet of water into a liquid cryomedium [144, 145]. Later, Mayer improved the technique by spraying micrometer-sized droplets on a solid cryoplate, thus avoiding the use of a cryomedium [146]. The droplets are sprayed into a vacuum chamber, which results in the formation of a supersonic jet of droplets. This jet of droplets hits a piece of copper cooled to 77 K, where the droplets are immobilized almost instantaneously. The deposit is called HGW.

### 3. *Preparation from the Solid Phase*

*3.1. High-Density Amorphous Ice (HDA).* When hexagonal ice (ice Ih) is pressurized melting can occur, e.g., at 253 K and 0.2 GPa, because one of water's anomalies is its negative melting volume, which results in a negatively sloped melting curve according to the Clausius-Clapeyron equation (c.f. Fig. 7). Extrapolating the ice Ih melting curve to lower temperatures, one would expect melting at pressures exceeding $\sim$1.0 GPa at 77 K or $\sim$ 0.6 GPa at 170 K if one could avoid recrystallization to high-pressure polymorphs of ice. Whereas at 170 K,

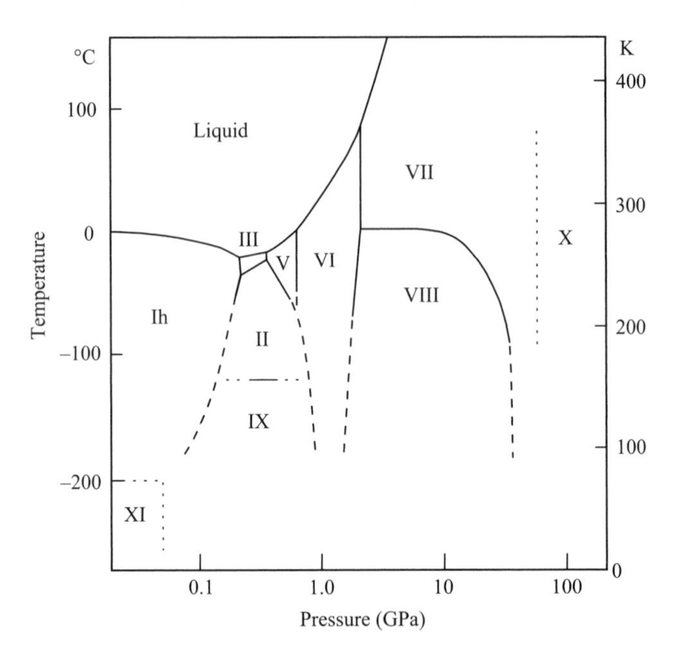


**Figure 7.** Phase diagram of water's stable crystal polymorphs. Metastable polymorphs such as ice IV or ice XII do not show up. Adapted from Ref. [147].

mixtures of stable ice II and metastable ice III/IX are indeed observed in experiment (c.f. Fig. 7) [148], at 77 K recrystallization is kinetically inhibited. Instead, pressure-induced amorphization takes place, and HDA forms after compressing hexagonal ice [149] or cubic ice [150] beyond 1.1 GPa. It is debated whether this observation can indeed be interpreted as "thermodynamic melting" immediately followed by vitrification of the liquid to the glass (HDA). Alternatively, it was suggested that overpressurization of the crystal results in a collapse of its lattice, i.e., "mechanical melting" producing "nanocrystallites" (see Section III.E) [151]. The density of this amorphous state at 77 K and 1 bar is 1.17 g/cm$^3$ [152].

*3.2. Low-Density Amorphous Ice (LDA).* Upon heating HDA to T > 115 K or very high density amorphous ice (VHDA) to T > 125 K at ambient pressure, the structurally distinct amorphous state LDA is produced. Alternatively, LDA can also be produced by decompressing HDA or VHDA in the narrow temperature range of 139–140 K to ambient pressure [153–155]. The density of this amorphous state at 77 K and 1 bar is 0.93 g/cm$^3$ [152]. These amorphous–amorphous transitions are discussed in Sections III.C and III.D.

As a third route to LDA, heating or decompression of the high-pressure polymorph ice VIII can be employed, which is the proton-ordered pendant to ice VII. Most high-pressure forms of ice quench-recovered at 77 K experience a transition to ice Ic when heated above 145–150 K, namely the metastable poymorphs ice IV and ice XII (which can be produced in the stability fields of ice V and ice VI in Fig. 7) [156] and the stable polymorphs ice II, ice III, ice V, ice VI, and ice VII [157, 158]. When ice VIII is heated at 1 bar, temperature-induced amorphization takes place, however [159]. A sequential transformation to first HDA and then LDA was observed when heating ice VIII at 1 bar [160]. Also, the direct route from ice VIII to LDA has been reported: Ice VIII decompressed to 1 bar at 80 K and then heated to 125 K directly transforms to LDA [161]. Also isothermal pathways to LDA by decompression are feasible: By decompressing partially hydrogen ordered ice VII' or ice VIII at 135 K, a transformation directly to LDA is observed using in situ Raman spectroscopy in a diamond anvil cell [162, 163].

Finally, radiation can be employed to produce LDA. Ice III or its proton-ordered counterpart ice IX are amorphized by particle bombardment at electron doses above 2400 electrons $nm^{-2}$ [164]. Ice I amorphizes by keV ion-bombardment at 10–80 K [165]. Similarly, after a dose of few eV per mol of UV photons amorphization of ice I is observed [166, 167]. The conversion rates increase as the temperature decreases [168]. Using 700 keV proton irradiation at 13 K, even oscillations between crystalline and amorphous ice can be achieved, whereas above 27 K, the amorphous ice remains [169].

*3.3. Very High Density Amorphous Ice (VHDA).* By annealing HDA to $T > 160$ K at pressures $> 0.8$ GPa, a state structurally distinct from HDA can be produced, which is called VHDA ice [152]. The structural change of HDA to a distinct state by pressure annealing was first noticed in 2001 [152]. Even though VHDA was produced in experiments prior to 2001 [170], the structural difference and the density difference of about 10% at 77 K, and 1 bar in comparison with HDA remained unnoticed. Powder X-ray diffraction, flotation, Raman spectroscopy, [152] neutron diffraction [171], and in situ densitometry [172, 173] were employed to show that VHDA is a structural state distinct from HDA. Alternatively, VHDA can be prepared by pressurization of LDA to $P > 1.1$ GPa at 125 K [173, 174] or by pressure-induced amorphization of hexagonal ice at temperatures $130\,K < T < 150\,K$ [170]. The density of this amorphous state at 77 K and 1 bar is 1.26 $g/cm^3$ [152].

## B. Structural Information

Structure factors and radial distribution functions (RDFs) were found to be nearly identical for ASW, HGW, and LDA by X-ray and neutron diffraction measurements [175]. A more recent isotope substitution neutron diffraction study on three sets of samples ($D_2O$, HDO, and $H_2O$) allowed determining the partial OO-, OH-, and HH-radial distribution functions [20]. As an example, the OO-RDF for ASW, HGW, and LDA is shown together with HDA and VHDA in

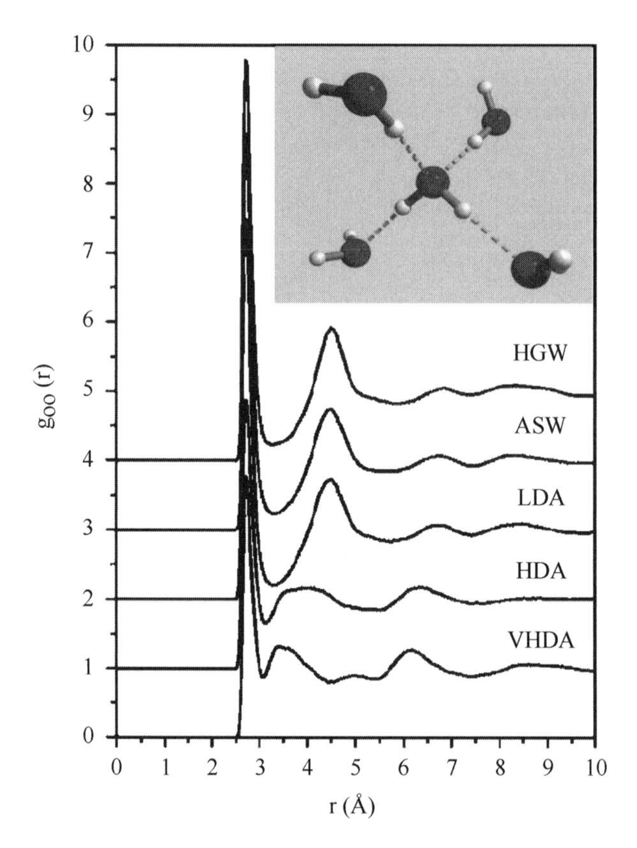


**Figure 8.** The EPSR estimated intermolecular oxygen–oxygen pair distribution function $g_{OO}(r)$. For clarity, the functions are offset by a unit. The inset shows the Walrafen pentamer, which is the structural motif common to all known amorphous ices. Adapted from Ref. [20]. See color insert.

Fig. 8. All the partial RDFs for ASW, HGW, and LDA agree within the experimental error, which suggests that ASW, HGW, and LDA all represent the same structural state at 77 K and 1 bar. The basic short-range order structural motif is the "Walrafen pentamer" (i.e., a central oxygen atom, which is surrounded tetrahedrally by four oxygen atoms; c.f. inset Fig. 8.). However, whereas their structures appear identical, it was suggested from inelastic incoherent neutron scattering that the dynamics of lattice and internal vibrations of water molecules differ significantly in HGW and LDA [176].

By contrast, HDA and VHDA differ substantially from ASW, HGW, and LDA in terms of local structure. Most importantly, there is an increased probability of finding water molecules at an OO-distance of 3.0–3.5 Å from the central water molecule. In case of HDA, one molecule is found at an interstitial position [177], and in the case of VHDA, two molecules occupy the interstitial positions [171]. That is, the coordination number increases from roughly 4 to 5 and 6 for LDA, HDA, and VHDA, respectively.

Information on the nearest-neighbor OO-distances was provided from nuclear magnetic resonance (NMR) measurements at 77 K, which resulted in 2.84 Å for HDA and 2.79 Å for LDA [178]. Raman measurements on $H_2O$ samples that contain 5 mol% $D_2O$ can be used to infer near-neighbor OO-distances by calling on a good correlation with the frequency of the decoupled OH stretching band. This correlation suggests OO-distances of 2.85 Å for VHDA, 2.82 Å for HDA, and 2.77 Å for LDA [152]. Thus, increasing density is coupled with increasing OD...O distance. This seemingly paradox situation has been attributed to an increase in coordination number [179].

LDA and HDA were interpreted to be similar to two limiting structural states of supercooled liquid water up to pressures of 0.6 GPa and down to 208 K. In this interpretation, the liquid structure at high pressure is nearly independent of temperature, and it is remarkably similar to the known structure of HDA. At a low pressure, the liquid structure approaches the structure of LDA as temperature decreases [180–182]. The hydrogen bond network in HDA is deformed strongly in a manner analogous to that found in water at high temperatures, whereas the pair correlation function of LDA is closer to that of supercooled water [183]. At ambient conditions, water was suggested to be a mixture of HDA-like and LDA-like states in an approximate proportion 2:3 [184–186].

### C. Irreversible Structural Transitions by Heating at 1 Bar

All amorphous ices transform to cubic ice when heated to $T > 150$ K at 1 bar, which subsequently recrystallizes to hexagonal ice. However, the high-density amorphous states HDA and VHDA also show an irreversible transition to LDA when heated to temperatures slightly below crystallization. Using ultrasonic sound propagation measurements, three distinct stages were found for the temperature-induced HDA→LDA transformation at 110–115 K, namely shear elastic softening, bulk softening, and main volume jump [187]. Whereas the first stages involve structural relaxation, the final stage is consistent with a "first-order like" transition—that is, the amorphous–amorphous transition shows a complex nonergodic nature that involves more than one process [65]. The relaxation stages of this combined process were studied in detail using neutron scattering [188, 189].

After heating VHDA at 1 bar to ~125–130 K, LDA is produced. There is only one exothermic peak in differential scanning calorimetry(DSC) measurements related to this transition [190]. However, elastic and inelastic neutron scattering experiments performed at narrowly spaced temperatures indicate that an amorphous structure indiscernible from HDA (produced by pressurizing ice Ih at 77 K) forms as an intermediate stage [191] The small angle scattering indicates that both VHDA and LDA are homogeneous, whereas the intermediate

HDA shows heterogeneities on a length scale of a few nanometers [192]. The activation energy for the VHDA→LDA transition is at least 20 kJ/mol higher than the activation energy for the HDA→LDA transition [191] (i.e., VHDA is thermally more stable at 1 bar than HDA) [170, 190]. Similarly, deuteron spin-lattice relaxation time T1-measurements for the VHDA→LDA transition suggest that an HDA-like state is incurred on the way. The HDA-like state on the way shows a much higher transition temperature to LDA than HDA (produced by pressurizing ice Ih at 77 K), which suggests that it is in a relaxed HDA state [193]. Mishima and Suzuki have performed Raman measurements on a VHDA sample and have monitored the transition to LDA as a function of temperature, time, and position on the amorphous sample [194]. In Fig. 9, it is shown that the measured VHDA Raman peak shift and intensity as well as the estimated sample density first continuously change toward HDA in the temperature range 100–115 K and then discontinuously change (marked by the arrows in Fig. 9) at ~ 116 K. Mishima and Suzuki [194] noticed that in the course of the transition, a "phase" boundary propagates through the sample. These results confirm a picture of a combined process that first involves structural relaxation at 1 bar followed by a "first-order like" transition that involves HDA-LDA coexistence. Please note, however, that these transitions are irreversible in the sense that going back to 77 K, the back transformation to HDA or VHDA, cannot be achieved. Thus, equality of Gibbs free energies is not involved in this case—rather, LDA has a much lower Gibbs free energy at 1 bar than HDA or VHDA.

### D. Reversible in situ Structural Transitions

By contrast, after changing the pressure, a reversible transition is observed. The upstroke transition from LDA to HDA was first observed by noticing a sharp 20% volume change at 0.55–0.65 GPa and 77 K, which causes a shift in the halo peak of the powder X-ray diffractogram of the quench-recovered state [25]. At higher temperatures, the sharp LDA→HDA transition shifts to lower pressures [153] (e.g., to 0.40–0.50 GPa at 125 K) [174]. Whereas HDA can be quench recovered at 77 K without structural change, it transforms back to LDA on the downstroke between 130 K and 140 K [153]. Under strictly isothermal conditions, the sharp downstroke transition is observable solely in a narrow temperature region. The transition is observed at 139 K and ~0.03 GPa [154] and at 140 K and ~0.06 GPa [155], whereas no such transition is observed at 136 K. Instead, HDA is recovered after decompressing to ambient pressure at 136 K. At 142 K, crystallization of HDA to an ice IX/ice V mixture is observed rather than the downstroke HDA→LDA transition. An observation of the HDA→LDA transition at temperatures $T < 136$ K would require negative pressure (i.e., a substantial hysteresis is involved).

When pressurizing LDA and by monitoring the structural state using in situ neutron diffraction, double-peaked diffraction patterns and a progressing

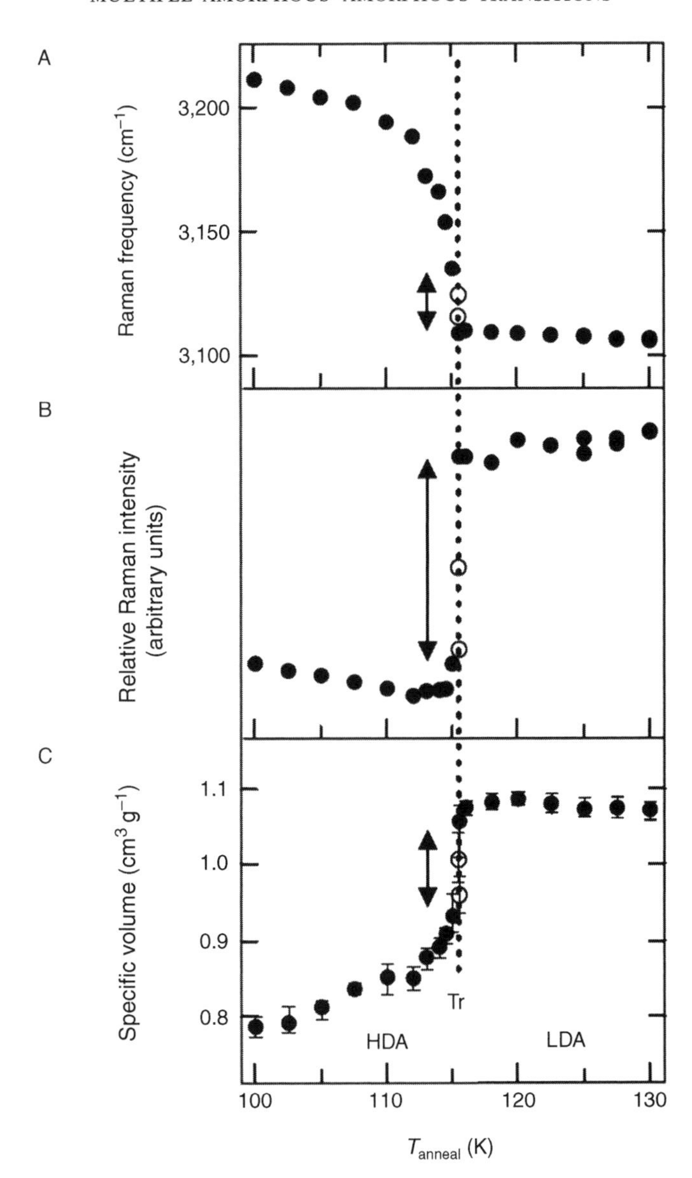


**Figure 9.** Changes in properties of VHDA ice induced by annealing at 1 bar. Up to 115 K continuous changes toward HDA can be observed. The discontinuous changes at the transformation at ~116 K are marked by gaps/arrows. Raman spectra can all be expressed as a superposition of pure LDA and pure HDA spectra. The specific volume in the bottom panel is estimated from visual inspection of the sample expansion. Reproduced from Ref. [193].

transformation to HDA were observed at ∼0.3 GPa and 130 K [195]. The intermediate double-peaked patterns shown in Fig. 10, panels (b)–(e), can be decomposed into a linear combination of the patterns of pure LDA and HDA, which suggests coexistence of the two amorphs. A progressive transformation from one state (LDA) to another state (HDA) is consistent with the data obtained during pressurization and is suggested by the authors to be evidence for a "classic first-order," transition. Please note, however, that there is an inherent none-quilibrium character to amorphous–amorphous transitions, because amorphous states by definition structurally relax on the time-scale of the experiment. This structural relaxation, even though it is very slow and so small that it cannot be resolved within their experimental setup, is superimposed on the first-order transition. Therefore, we prefer the notion of a "first-order like," a "quasi first-order," or an "apparently first-order" transition, which was first used by Mishima et al. [25]. We also note that the conclusion of coexistence was doubted, and the occurrence of intermediate amorphous ices was suggested because (1) neutrons rather than X-rays were employed, (2) pressure gradients may blur the analysis, and (3) a peak shift parameter was necessary in the fit procedure [197].

We now turn from the single, possibly "first-order like" nature of the LDA–HDA transition to the multiple amorphous–amorphous transitions in water. When pressurizing LDA at slow rates at 125 K, a stepwise transition LDA→HDA→VHDA is observed [173]. The piston displacement and density data as obtained from a piston-cylinder experiment that shows the stepwise nature are depicted in Fig. 11. The upstroke HDA→VHDA densification of 5% takes place in the pressure range 0.80–0.95 GPa and is somewhat less sharp when comparing with the upstroke LDA→HDA densification of 20% at 0.40–0.50 GPa. During fast compression, the LDA→HDA transition is still sharp, whereas the HDA→VHDA transition is smeared out over a broad pressure range [173]. These findings support the possibility of an LDA–HDA first-order like transition but leave the question open whether a first-order like transition underlies the HDA→VHDA transition and/or whether the observed second step

**Figure 10.** Neutron diffraction patterns as LDA (top, $\eta = 1$) transforms to HDA (bottom, $\eta = 0$) at 130 K. The measured data are shown using open circles. Parameterizations to the data are shown using lines. Parameterizations of pure LDA at 0 GPa [panel (a)] and pure HDA at 0.5 GPa [panel (f)] are shown in the inset. The measured data on the upstroke transition at a roughly estimated pressure of ∼0.3 GPa [panels (b)–(e)] can be fitted to linear combinations of pure LDA and pure HDA patterns, where $\eta$ indicates the fraction of LDA in the binary LDA–HDA mixture. Peak shift parameters were employed to account for the peak shift incurred by bringing LDA from 0.0 GPa to 0.3 GPa and HDA from 0.5 GPa to 0.3 GPa. The subtraction of the fitted data from the measured data is shown below each pattern. The asterisk in panel (f) marks a peak of a small amount of untransformed hexagonal ice. The amount of the ice contamination remains constant during the whole experiment because the amorphization of hexagonal ice at 130 K requires pressures much higher than 0.5 GPa [174, 196]. Reproduced from Ref. [195].

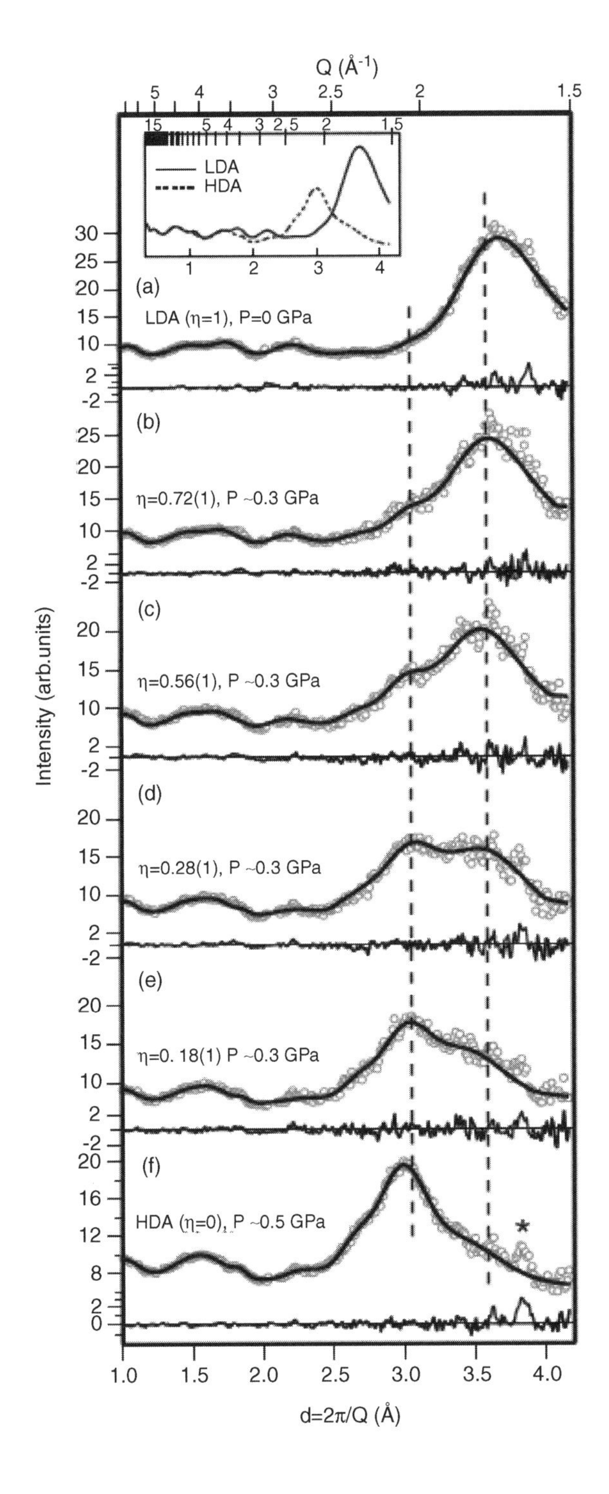

Q (Å⁻¹)
LDA
HDA
(a)
LDA (η=1), P=0 GPa
(b)
η=0.72(1), P ~0.3 GPa
(c)
η=0.56(1), P ~0.3 GPa
(d)
η=0.28(1), P ~0.3 GPa
(e)
η=0. 18(1) P ~0.3 GPa
(f)
HDA (η=0), P ~0.5 GPa
Intensity (arb.units)
d=2π/Q (Å)

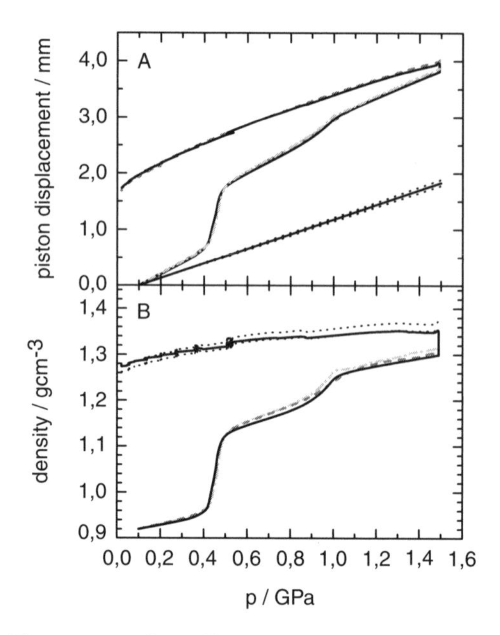


**Figure 11.** (a) Three curves of "raw" piston displacement data obtained on compressing 300 mg of LDA at 125 K in a piston-cylinder apparatus with a bore diameter of 8 mm at a rate of 20 MPa/min, quenching to 77 K at 1.5 GPa and subsequent pressure release at 77 K (top curves) together with the apparatus correction (straight line at bottom). (b) Density data calculated from the "raw" piston displacement curve and the apparatus correction with the use of Eq. (1) given in Ref. [174]. Adapted from Ref. [174].

is caused by a slow relaxation of HDA to the metastable equilibrium state at 125 K. One way of answering this question is by accelerating the relaxation dynamics by working at higher temperatures (i.e., by working as close as possible to the metastable equilibrium). The highest possible temperature at which crystallization does not interfere is 140 K in case of the downstroke transition [154]. When decompressing VHDA at 140 K, a single density step is observed at ~0.06 GPa, which is caused by the HDA→LDA downstroke transition [154]. No density step is related to the VHDA→HDA transition. Whereas the structural state of VHDA does not change in the pressure range from 1.1 GPa down to ~0.4 GPa, a continuous structural transition to HDA is observed in the range from ~0.40 GPa down to ~0.06 GPa. The corresponding X-ray diffractograms are shown in Fig. 12 [panel (a)] together with an analysis of the peak position [panel (b)] and full width at half maximum [panel (c)] for the first broad diffraction peak. Many intermediate states exsist between HDA and VHDA, which can be quench-recovered, whereas no intermediate states can be quench-recovered between LDA and HDA. The HDA state quench-recovered at 0.7 GPa in the upstroke transition at 125 K (cf. Fig. 11) resembles state

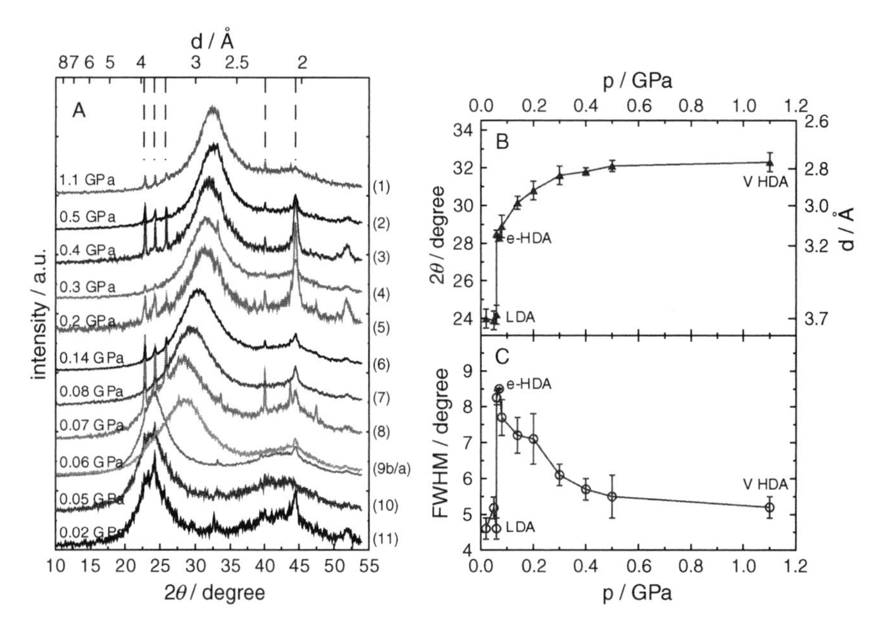


**Figure 12.** (a) Powder X-ray diffractograms obtained by isothermal decompression of VHDA (sample 1) at 140 K to several selected pressures, subsequent quenching to 77 K and recovery to ambient pressure (samples 2–11). The diffractograms (Cu K$\alpha$) were recorded at $\sim$83 K on a diffractometer in $\Theta$–$\Theta$ geometry (Siemens, New York, NY; model D 5000), equipped with a low-temperature camera of Anton Paar. Curves are shown on the same scale and smoothed with a 21-point Savitzky-Golay polynomial of fifth order. Data are offset and scaled for clarity. The right panel shows (b) the peak position and (c) the full width at half maximum (FWHM) of the first broad diffraction peak as a function of the selected decompression pressure. For part (b), the experimental error of a single measurement is $\pm$ 0.5°. In some powder diffractograms, sharp features marked by the dashed lines can be observed. They develop from traces of $I_h$, which had formed by condensation of water vapor during transfer of the sample onto the precooled sample holder [198], the X-ray sample holder itself (chrome-plated Cu: $2\Theta = 44.5°$), or remnants of indium. The samples themselves are fully amorphous. Reproduced from Ref. [155].

number 6 in Fig. 12, which was quench-recovered on the downstroke transition from 140 K and 0.14 GPa. The occurrence of intermediate states is evidence that no first-order character underlies the HDA$\leftrightarrow$VHDA transition, even though both the upstroke and the downstroke transition take place in a narrow pressure interval. We note that the HDA, which transforms to LDA in a first-order-like manner, is a structural state distinct from the HDA, which is produced by pressure-induced amorphization of hexagonal ice at 77 K, because the latter is unrelaxed and far away from the metastable equilibrium state, whereas the former is structurally relaxed and much closer to the metastable equilibrium. Nelmes et al. [199], therefore, suggested distinguishing the two states by calling the latter "unrelaxed HDA" (uHDA) and the former "expanded HDA" (eHDA).

The density of eHDA is 1.13 g/cm$^3$, which is about 4% less than the density of uHDA [200]. The different degree of structural relaxation should be evident when studying the thermal stability at 1 bar. The well-relaxed state is expected to be thermally more stable, and this expectation is confirmed experimentally: At 1 bar, eHDA is stable up to ~135 K, whereas uHDA is stable merely up to ~115 K [190, 194, 196]. Earlier, Johari [201] had noticed that HDA produced from LDA ("eHDA") shows ultrasonic properties that differ from HDA produced from ice Ih ("uHDA").

A summary in the form of a "diagram of metastable amorphous states" is depicted in Fig. 13, in which the upstroke–downstroke hysteresis has been averaged out. The vertical arrow represents the pressure annealing from

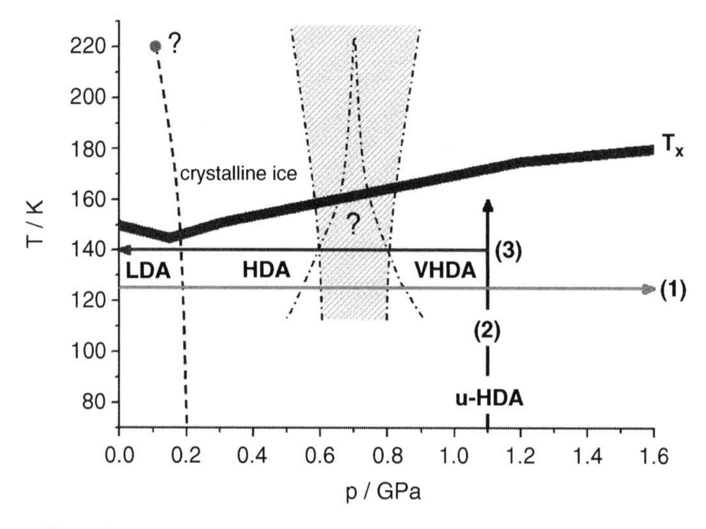


**Figure 13.** Schematic phase diagram of water's metastable states. Line (1) indicates the upstroke transition LDA→HDA→VHDA discussed in Refs. [173, 174]. Line (2) indicates the standard preparation procedure of VHDA (annealing of uHDA to 160 K at 1.1 GPa) as discussed in Ref. [152]. Line (3) indicates the reverse downstroke transition VHDA→HDA→LDA as discussed in Ref. [155]. The thick gray line marked $T_x$ represents the crystallization temperature above which rapid crystallization is observed (adapted from Mishima [153]). The metastability fields for LDA and HDA are delineated by a sharp line, which is the possible extension of a first-order liquid–liquid transition ending in a hypothesized second critical point. The metastability fields for HDA and VHDA are delineated by a broad area, which may either become broader (according to the singularity free scenario [202, 203]) or alternatively become more narrow (in case the transition is limited by kinetics) as the temperature is increased. The question marks indicate that the extrapolation of the abrupt LDA↔HDA and the smeared HDA↔VHDA transitions at 140 K to higher temperatures are speculative. For simplicity, we average out the hysteresis effect observed during upstroke and downstroke transitions as previously done by Mishima [153], which results in a HDA↔VHDA transition at T = 140 K and P ~ 0.70 GPa, which is ~ 0.25 GPa broad and a LDA↔HDA transition at T = 140 K and P ~ 0.20 GPa, which is at most 0.01 GPa broad (i.e., discontinuous) within the experimental resolution.

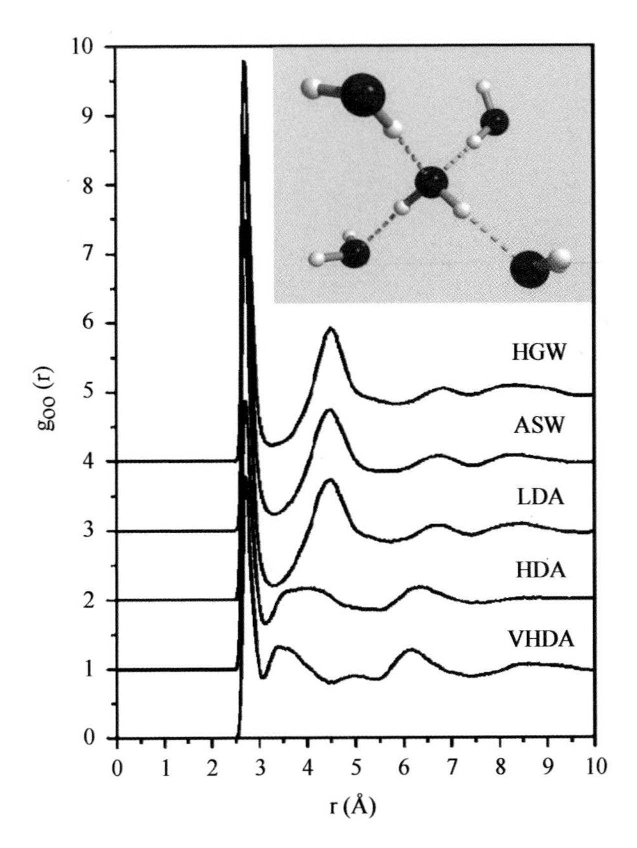


**Figure 8.** The EPSR estimated intermolecular oxygen–oxygen pair distribution function $g_{OO}(r)$. For clarity, the functions are offset by a unit. The inset shows the Walrafen pentamer, which is the structural motif common to all known amorphous ices. Adapted from Ref. [20].

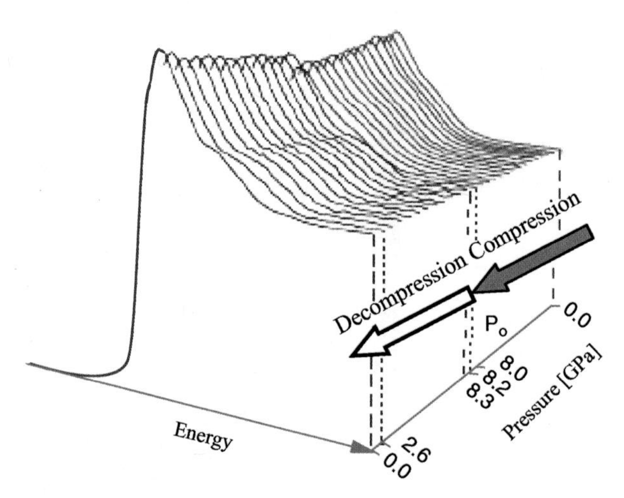


**Figure 21.** X-ray absorption spectroscopy data of a-Ge in a compression–decompression cycle [272].

unrelaxed HDA (produced at 77 K from ice Ih) at 1.1 GPa to 160 K to VHDA. VHDA represents the ultimately densified amorphous state of water, which crystallizes rather than densifies when compressed beyond 2 GPa [204]. The left-facing arrow shows the downstroke transition at 140 K. VHDA gradually transforms to HDA in a finite pressure interval. To distinguish this relaxed form of HDA from uHDA, the term "expanded HDA" is preferable. eHDA represents the ultimately expanded state of HDA, which transforms to LDA possibly in a first-order like transition and which may coexist with LDA. Such a coexistence has been inferred for the upstroke transition [195]. Continued compression of HDA along the right-facing arrow results in a second transition to VHDA, which is continuous but confined to a finite pressure interval. Both eHDA and VHDA are well relaxed and at least close to being metastable equilibrium states. uHDA, however, is an instable state and inevitably relaxes structurally during heating. As a working hypothesis, one can assume that eHDA is the closest possible amorphous state directly related to a deeply supercooled high-density liquid on the low pressure side (e.g., at 0.2 GPa) and that VHDA is the closest possible proxy of a deeply supercooled very high-density liquid on the high-pressure side (e.g., at 1.5 GPa). In addition, there is a possible low-density liquid related to LDA, and hence, there may be three deeply supercooled liquids. A direct experimental proof for such a correspondence between amorphous states and deeply supercooled liquids is missing currently. What needs to be done here is to measure the in situ relaxation times directly and to determine whether a glass → liquid transition occurs during isobaric heating. The first results obtained by dielectric relaxation spectroscopy of VHDA at 1 GPa suggest that VHDA indeed shows liquid-like relaxation behavior at 140 K [205, 206]. The question of whether the amorphous states of water behave like glasses (i.e., vitrified liquids), which turn into liquids during heating, or whether they behave like crystals, which do not liquify during heating, is highly disputed as outlined in the following section.

### E. Are the Amorphous Solids Glasses or Nanocrystallites?

The question of whether there is a true glassy nature of amorphous ices is of interest when speculating about possible liquid–liquid transitions in (deeply) supercooled water. For true glasses, the amorphous–amorphous transitions described here can be viewed as the low-temperature extension of liquid–liquid transitions among LDL, HDL, and possibly VHDL. That is, the first-order like LDA ↔ HDA transition may map into a first-order LDL ↔ HDL transition, and the continuous HDA ↔ VHDA transition may map into a smeared HDL ↔ VHDL transition. Many possible scenarios are used how to explain water's anomalies [40], which share the feature of a liquid–liquid transition [202, 207–212]. They differ, however, in the details of the nature of the liquid–liquid transition: Is it continuous or discontinuous? Does it end in a liquid–liquid critical point or at the reentrant gas–liquid spinodal?

Except for HGW, which is commonly accepted to be a true glass, there is a lot of debate and scientific discourse regarding a possible nanocrystalline nature of amorphous ices. Evidence is abundant in the literature both for a "pro-glass" and a "contra-glass" view.

In the case of LDA, it has been suggested that it behaves in many respects similar to ice I. Comparison of vibrational spectra and oxygen *K*-edge X-ray absorption spectra of LDA with ice [151, 213, 214] and an analysis of Gibbs free energies (which indicates that LDA would need to transform to ice Ih by pressurizing to 0.7 GPa at 77 K rather than HDA if LDA were indeed a glass) suggest that LDA is not a truly amorphous material [215]. The absence of fast precursor dynamics in LDA during heating from 2 K to 170 K in elastic high-Q backscattering neutron diffraction was interpreted as evidence against the glassy nature of LDA, because fast precursor dynamics would be expected from predictions by mode coupling theory near the glass transition [216]. Inelastic X-ray scattering in the 1–15 $nm^{-1}$ momentum transfer (Q) range shows sharp, crystal-like phonons [217]. Also, crystal-like features in the thermal conductivity of LDA indicate that phonon–phonon scattering is dominant [218]. LDA, like cubic ice and hexagonal ice, shows a negative Bridgman parameter, which characterizes the density dependence of the thermal conductivity. However, the 1H-spin lattice relaxation time of LDA (and also HDA) at 77 K is very different from the 1H-spin lattice relaxation time of crystalline ices [178]. Also, a thermally reversible glass$\rightarrow$liquid transition was measured in LDA at a temperature of 134 K, which is accompanied by a change in heat capacity $\Delta c_p \sim 0.7$ J/Kmol [219]. (Please note that $T_g$ for LDA was initially erroneously reported to be 129 K [220–222]). The DSC curves for LDA are remarkably similar to the DSC curves of annealed ASW [138] and HGW [223–226], which both show a $T_g$ of 136 K at a heating rate of 30 K/min. The activation energy of structural relaxation near $T_g$ was determined to be $\sim$55 kJ/mol [224]. So, judging from DSC experiments, HGW, ASW, and LDA are very similar not only in terms of static structure as outlined in Section III.B, but also in terms of their glass$\rightarrow$liquid transition [138, 219]. Ultrafast scanning of pure ASW samples at a heating rate of up to $10^5$ K/s seems to be in contradiction, because no sign of a glass$\rightarrow$liquid transition up to $\sim$205 K was reported, where it crystallizes to ice Ic [227]. However, Johari [228] suggested that this conclusion was made prematurely, and a reanalysis of the data is necessary. To resolve the question of whether a difference exists between LDA, which may be nanocrystalline or a glass, and HGW, which is definitely a glass, it would be highly instructive to study vibrational spectra and thermal conductivity in HGW samples and compare the results directly with the LDA case.

In the case of ASW, the self-diffusivity shows contradicting experimental evidence. At temperatures up to 125 K, the mobility of protons injected into annealed $D_2O$-ASW at 80 K was shown to be consistent with orientational diffusion based on L-defect activity rather than molecular diffusion. This finding suggests that fluidity does not develop when heating to above the glass→liquid transition at 136 K and that the isotopic exchange behavior in annealed $D_2O$-ASW resembles the behavior in cubic ice [229]. However, the self-diffusivity as measured by hydrogen/deuterium (H/D) isotope exchange in the 150–160 K range is roughly a millionfold greater than that expected for crystalline ice and is in accordance with a fluid-like translational diffusive motion. This is consistent with an amorphous solid that melts into a deeply supercooled liquid prior to crystallization [230, 231]. Self-diffusivity measurements performed by thermal desorption spectroscopy in layered films of ASW and organic spacers indicate that interlayer mixing in the 150–160 K range does not occur by diffusion through a dense phase but through an interconnected network of cracks/fractures created within the ASW film during crystallization. This implies that the self-diffusivity of ASW below crystallization is inconsistent with a "fragile" liquid and leaves the two options of (1) $T_g > 160$ K or (2) the liquid above $T_g \sim 136$ K is a "strong" liquid [232, 233]. The picture of ASW transforming into a "strong" liquid above $T_g \sim 136$ K is supported by electron diffraction studies of the crystallization behavior of ASW. These studies show an onset of the amorphous relaxation (coincident with the glass transition) *prior* to crystallization. Above the glass transition temperature, the crystallization kinetics, film morphology changes, changes in binding energy of water molecules, and band shape changes are consistent with the amorphous solid becoming a "strong" ultraviscous liquid [234, 235]. The concept of a dynamic crossover from "fragile" (at $T > 230$ K) to "strong" (at $T < 150$ K) was also invoked for the case of HGW. Although it was suggested initially that the weak glass→liquid transition in HGW at ~136 K [222] would merely be a shadow of a pronounced but hidden glass→liquid transition at ~160–165 K [236, 237], the consensus is now that the weak endothermic step observed at ~136 K is a real glass→liquid transition to a deeply supercooled "strong" liquid [212, 225, 226, 238].

In the case of HDA, it has been proposed that HDA may be a mixture of highly strained nanocrystalline high-pressure phases of ice instead of being a homogeneously random structure [239]. Also for VHDA, it was suggested [240] that its structure factor is highly reminiscent of the structure factor obtained for a mechanically collapsed and densified ice [241]. From inelastic neutron scattering, it was inferred that HDA shows vibrational spectra similar to ice VI [242], (i.e., short-range atomic correlations and force constants are similar in HDA and ice VI, whereas the degree of disorder on a long-range scale differs). Inelastic neutron scattering (INS) shows that the first phonon peak in the 0.5–20 meV range is softer for LDA compared with HDA [243]. However, INS in the energy transfer region 2–500 meV (i.e., 16–4025 $cm^{-1}$) shows HDA to behave glass-like in the translational and librational regions (<150 meV) [214]. Other INS studies illustrate clearly an excess number of modes in the HDA density of states at 5 K centered at 0.65 THz, which is not found in LDA, ice Ic, or ice Ih [244]. The thermal conductivity of HDA ice under pressure, by contrast to LDA, follows the behavior expected for a glass and a positive Bridgman parameter [218]. Similarly, the Gruneisen parameter that characterizes low-frequency phonons is negative for LDA (i.e., crystal like), but is positive for HDA (i.e., glass like) [245]. Also, the two-level system density of states in HDA is comparable with that found in many conventional glasses (by contrast to LDA) [246].

The phonon dispersion of hexagonal ice measured by inelastic neutron scattering up to 0.5 GPa at 140 K reveals a pronounced softening (e.g., for a transverse acoustic phonon branch), which is suggested to be at the origin of anomalous features of hexagonal ice, such as its negative thermal expansion coefficient below 60 K and solid-state amorphization [247]. Extrapolation of the data to 2.5 GPa, where some mode frequencies approach zero, suggests that pressure-induced amorphization of hexagonal ice is caused by mechanical melting rather than by thermodynamic melting. An ultrasonic study suggests that the ice Ih $\rightarrow$ HDA transition is, very much alike the LDA $\rightarrow$ HDA transition, preceded by elastic softening. This finding can be interpreted in favor of the crystal lattice instability paradigm [187, 248]. Whereas at low temperatures ($<162$ K), the Born stability criterion of lattices is violated ("mechanical melting"), at higher temperatures ($>162$ K) a Lindemann transition is observed ("thermodynamic melting") [151, 249]. It has been noted that the mechanism of solid-state amorphization is not only temperature dependent but also time- and pressure-dependent, and it cannot be described in terms of Born/Lindemann criteria as long as crystal-size effects (stresses at grain boundaries, etc.) and production of lattice faults during uniaxial pressurization are incorporated properly [239, 250].

The best evidence so far for the glassy nature of HDA was provided (1) by measurements of the dielectric relaxation time under pressure at 140 K [206, 251], (2) by the direct vitrification of a pressurized liquid water emulsion to HDA [252], and (3) by a high-pressure study of the glass $\rightarrow$ liquid transition using differential thermal analysis (DTA) [253]. We note here that these studies probe structurally relaxed HDA (eHDA) rather than unrelaxed HDA. It is possible that structurally relaxed HDA behaves glass like, whereas structurally uHDA shows a distinct behavior. Thus, more studies are needed in the future, which directly compare structurally relaxed and unrelaxed HDA.

## IV. SEMICONDUCTORS

### A. General Comments

The tetrahedral open network is a specific characteristic not only of water and silica but also of covalent systems such as Si and Ge (group IV semiconductors) [254, 255]. These substances share many characteristics with water, such as locally tetrahedral coordination at ambient pressure, negative melting slope in the P-T phase diagram, denser liquid than its crystalline form, and so on. All of these characteristics come from the tetrahedral network structure, which in fact plays a key role in water's polyamorphism. Because the tetrahedral network in Si and Ge is preserved in amorphous forms as well as in crystalline forms at ambient pressure, one naturally expects that Si and Ge exhibit polyamorphic transformations similar to those in water (ice).

Unlike that in water, the tetrahedral network in Si and Ge mainly consists of covalent interatomic bonding, which makes the network more rigid than hydrogen bonding, leading to relatively high melting temperatures (e.g., 1687 K for *c*-Si). Although crystalline and amorphous states at ambient pressure hold the rigid network that exhibits a semiconducting nature, liquid and solid states under high pressure contain a highly distorted or collapsed tetrahedral network, which results in a metallic nature. The semiconductor–metal transition would thus be likely to accompany polyamorphic transformations in Si and Ge, in contrast to those in water. Because of such a substantial change in the electronic state (interatomic bonding), a two-state model [256–259] has often been invoked to describe polyamorphic transformations in Si and Ge. This model was originally developed by Rapoport [256] to account for the anomalous (negative) slope of the melting curve in tetrahedrally coordinated liquids.

## B. Amorphous Si

Convincing evidence for multiple amorphous phases in Si has been provided only recently. The first indication of pressure-induced amorphous–amorphous transition in Si was reported by Shimomura and colleagues [260, 261]. The electronic resistance of an amorphous Si (*a*-Si) film of about 1 μm thickness was measured in a compression–decompression cycle [260]. The electronic resistance showed a substantial drop at about 10 GPa in the compression process, which indicates a transition to a metallic state. In the decompression process, in contrast, the electronic resistance showed a large hysteresis and returned to the original semiconducting value only gradually. X-ray diffraction patterns were measured at 0 GPa for both the initial and final samples in the compression–decompression cycle, and amorphous patterns were obtained for both samples [260]. However, the X-ray diffraction patterns were not obtained for the metallic sample under pressure, so it remains uncertain whether densified metallic Si retained an amorphous form at ∼10 GPa. In fact, other experiments have shown that *a*-Si is transformed to β-tin crystal (a high-pressure polymorph [255]) above 10 GPa [262], which may have been realized in the experiment of Shimomura et al. [260].

Instead of pressurizing *a*-Si, Deb et al. [263] obtained a densified *a*-Si via pressurizing porous Si. They prepared films of porous Si having crystallite of ∼5 nm (on average). In situ measurements of X-ray diffraction patterns and

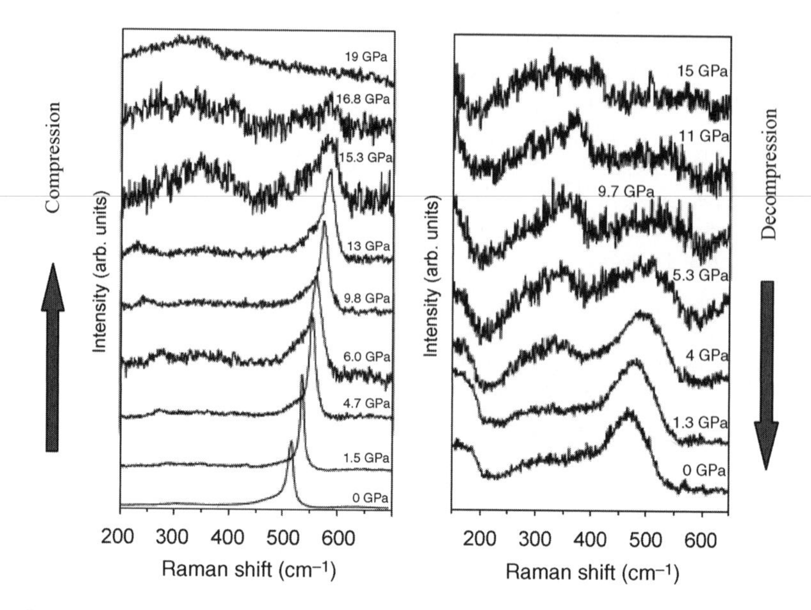


**Figure 14.** Raman spectra of porous Si in a compression–decompression cycle [263]. In the compression process, the characteristic spectrum of nanocrystalline Si disappears above ~13 GPa and a broad amorphous feature emerges. In the decompression process, the characteristic spectrum of the LDA form grows below ~9 GPa, which indicates an HDA-to-LDA transition.

Raman spectra for the sample were conducted in a compression–decompression cycle. In this experiment, the crystalline diffraction began to disappear above 7–8 GPa during compression, and pressure-induced amorphization was indicated by the Raman spectra above ~13 GPa (Fig. 14). The resultant HDA Si exhibits the Raman spectrum that differs from the spectrum of normal *a*-Si (LDA Si). Rather, the characteristics of the spectrum for HDA Si resemble those of the β-tin crystal, which indicates that HDA Si has a (locally) analogous structure to the β-tin structure. The synthesis of the HDA form of Si by Deb et al. [263] has a strong resemblance to that of water (ice) by Mishima et al. [149, 196]. Whereas compression induced amorphization that was almost completed at 13–15 GPa, decompression induced an HDA–LDA transition below 10 GPa, which is clearly shown in the Raman spectra (Fig. 14). This is the first direct observation of an amorphous–amorphous transition in Si. The spectrum at 0 GPa after the pressure release exhibits the characteristic bands of tetrahedrally coordinated *a*-Si (LDA Si). Based on their experimental findings Deb et al. [263] discussed the possible existence of liquid–liquid transition in Si by invoking a bond-excitation model [258, 259]. They have predicted a first-order transition between high-density liquid (HDL) and low-density liquid

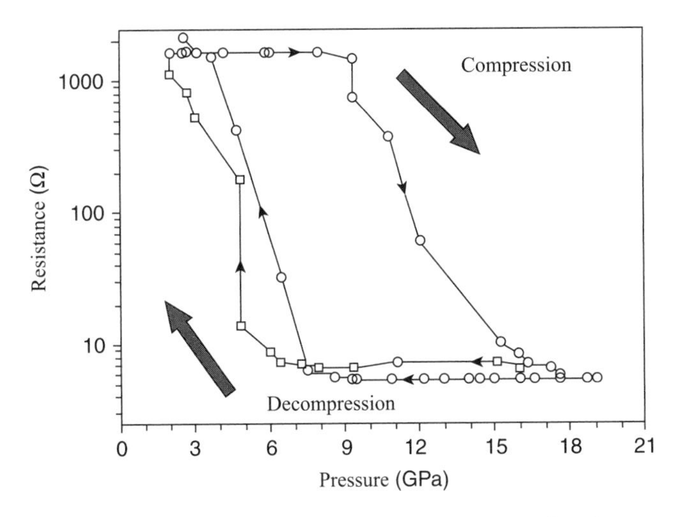


**Figure 15.** Electrical resistance measurements of a-Si in a compression–decompression cycle [264].

(LDL) below 1600 K at 1 atm and a second critical point at a negative pressure [263].

The transition from LDA to HDA Si was observed in the successive experiment by McMillan et al. [264]. In situ Raman spectra and electronic resistance measurements were performed with optical observation. After compression, the LDA form prepared by solid-state metathesis synthesis [10] was found to be transformed to the HDA form at ~14 GPa. The electronic resistance exhibited a sharp decrease at 10–14 GPa (Fig. 15), which is consistent with the early experimental findings by Shimomura et al. [260]. Optical micrographs show that HDA Si is highly reflective, whereas LDA Si is dark colored and nonreflective. This finding again supports that the LDA–HDA transition of Si is accompanied by a semiconductor–metal transition. Reverse transitions with large hysteresis were also observed; LDA Si began to form from HDA Si at 4–6 GPa after decompression from ~20 GPa.

It should be noted that no direct structural information about HDA Si was obtained in the experiments by Deb et al. [263] and McMillan et al. [264]. Their experimental data imply that HDA Si is structurally based on the β-tin structure, but the data are not sufficient to experimentally determine the structure of HDA Si.

Computer simulation studies have taken the lead in disclosing structural properties of HDA Si. In particular, ab initio calculations are playing a significant role in predicting the structural properties of HDA Si [265, 266]. Ab initio molecular-dynamics (MD) simulations based on plane-wave density functional theory (DFT) were performed by Morishita [265] to investigate the

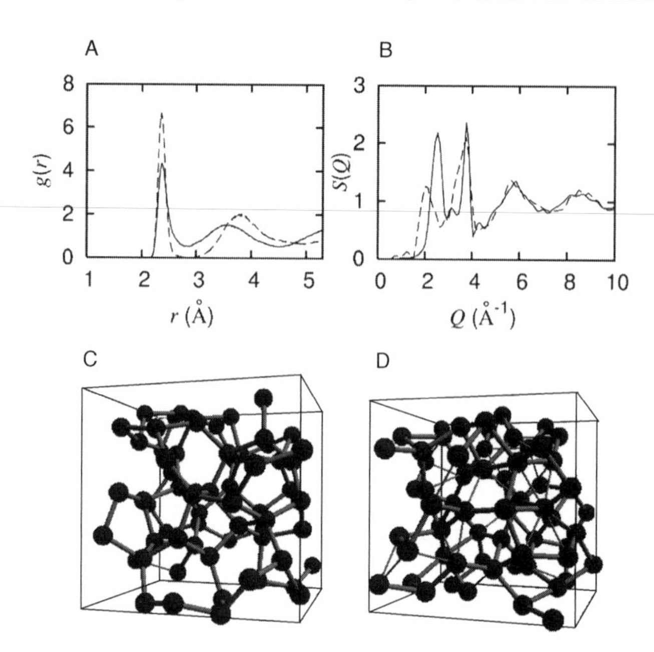


**Figure 16.** Structural profiles of LDA and HDA Si obtained from the plane-wave DFT calculations [265]. (a) Pair correlation functions g(r) for HDA (at 12 GPa, solid lines) and LDA (at 0 GPa, dashed lines). (b) Structure factors $S(Q)$ for HDA (at 12 GPa, solid lines) and LDA (at 0 GPa, dashed lines). (c, d) Atomic configurations for (c) LDA and (d) HDA. Atoms separated by 2.55 Å or less are linked by thick lines (covalent-like bonds), whereas those separated by 2.857 Å or less are linked by thin lines.

LDA–HDA transition of Si at an atomistic level. LDA Si was pressurized in a stepwise manner, whereas the temperature was maintained at 300 K. At 12 GPa, the LDA form was transformed to the HDA form with large volume reduction (~10%). The pair correlation functions g(r) for the LDA form and the HDA form thus obtained are given in Fig. 16. The first peak of g(r) is considerably broadened and becomes less intense after the transition to HDA. The second peak shifts to smaller *r*, and the separation between the first and second peaks is not as clear in HDA as in LDA. The coordination number $N_c$ obtained by integrating $4\pi r^2 \rho^* g(r)$ up to the first minimum is 4.0 for LDA, whereas it is 5.1 for HDA (where $\rho^*$ is the number density). Detailed structural analyses indicate that the first four neighboring atoms in HDA still form a (distorted) tetrahedral configuration, but the fifth neighboring atom is located at an open space of the tetrahedron (Fig. 16). It is thus considered that the HDA structure is constructed by forcing the fifth neighboring atom, which is outside of the first coordination shell of the LDA structure, into an interstitial position. This mechanism

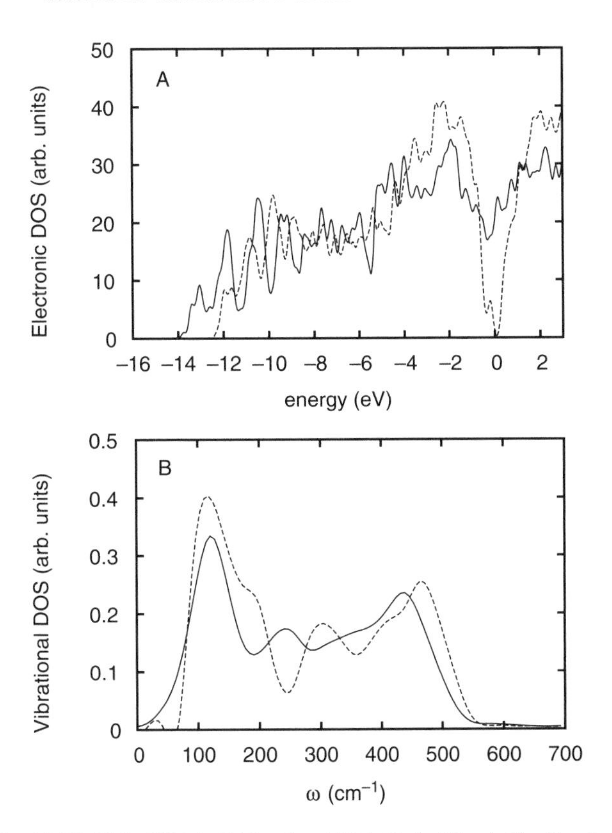


**Figure 17.** (a) EDOS for HDA (solid lines) and LDA (dashed lines). The Fermi energy is set to 0 eV. (b) VDOS for HDA (solid lines) and LDA (dashed lines). Both EDOS and VDOS were obtained from the plane-wave DFT calculations.

resembles the formation mechanism of the HDA ice structure [177], which indicates an inherent nature of tetrahedrally coordinated materials under pressure.

The electronic densities of states (EDOS) calculated for LDA and HDA Si (Fig. 17) confirm that HDA is metallic, as suggested by the experimental results [264]. There is no gap at the Fermi energy in EDOS for HDA, as Fig. 17 shows clearly. The calculated vibrational densities of states (VDOS) are also consistent with the previous experimental results [263, 264]. The LA and LO bands ($\sim$300 and $\sim$420 $cm^{-1}$, respectively) in LDA are broadened, and the TO band ($\sim$500 $cm^{-1}$) shifts to a lower frequency after the transition to HDA. This results in broad intensity in the range 200–450 $cm^{-1}$ in VDOS for HDA (Fig. 17). The overall profile is consistent with previous experimental findings [263, 264].

HDA Si obtained in Morishita's simulation shares traits with the β-tin form in the short-range atomic configuration. Both forms contain distorted tetrahedral configurations with $N_c$ of 5–6. However, ab initio calculations by Durandurdu and Drabold [266] demonstrated that other types of high-density amorphous Si could be formed by densification. On the basis of DFT calculation with a local-orbital basis set, they simulated a dynamical relaxation process from 800 K to 0 K for the LDA form at various pressures in the range of 0–17 GPa. The LDA form was preserved up to ~16 GPa, but it relaxed to a completely different amorphous form at a slightly higher pressure (16.25 GPa). The density of the resultant amorphous form is over 20% higher than that of the LDA form, and it is still higher than that of the HDA form obtained in Morishita's calculation [265]. We thus call this amorphous form VHDA Si to echo the terminology used to describe VHDA ice [152, 171, 265]. The characteristics of g(r) for the VHDA form (Fig. 18) are similar to those of the HDA form. However, the $N_c$ of VHDA is 8–9, which is much higher than that of HDA. The atomic configurations in VHDA, in fact, seem to contain fragments of the simple-hexagonal (sh) structure whose $N_c$ is 8. Note that the sh structure is experimentally obtained by pressurizing the β-tin structure of Si above ~16 GPa. [255]. VHDA Si is also found to be metallic and to have a relatively structureless profile of VDOS (Fig. 18).

These theoretical studies suggest the possibility of multiple amorphous–amorphous transitions in Si like those observed in water, although this has not yet been confirmed experimentally (Fig. 19). The notable point is that the pressure-induced sequence of amorphous Si (LDA, HDA, and VHDA) resembles that of crystalline Si (the diamond, β-tin, and sh structures [255]). This similarity suggests that amorphous structures are formed based on the corresponding crystalline structures according to the external pressure. We would like to point out, however, that many metastable crystalline structures are observed in Si [255]. This means that the free energy landscape is immensely complicated, and many local minima, each corresponding to an amorphous or crystalline form, are located in the landscape. It is therefore likely that many variants of the HDA or VHDA forms can be observed experimentally, depending on the sample preparation as well as on the external conditions (temperature and pressure).

Very recently, the structure of HDA-Si was experimentally disclosed by means of X-ray diffraction measurements [267, 268]. Daisenberger et al. [267] have attempted to obtain the structure factor $S(Q)$ for the HDA form under pressure. Figure 20 shows the experimentally obtained $S(Q)$ for LDA (3–13.5 GPa) and HDA (16.5 GPa). As the pressure is increased from 13.5 to 16.5 GPa, the first peak becomes more intense than the second peak and shifts to larger $Q$. The shift of the first peak indicates a large densification, and the overall features are in good agreement with the ab initio MD results [265] (Fig.16). Unfortunately, however, g(r) calculated from the $S(Q)$ by Fourier transformation

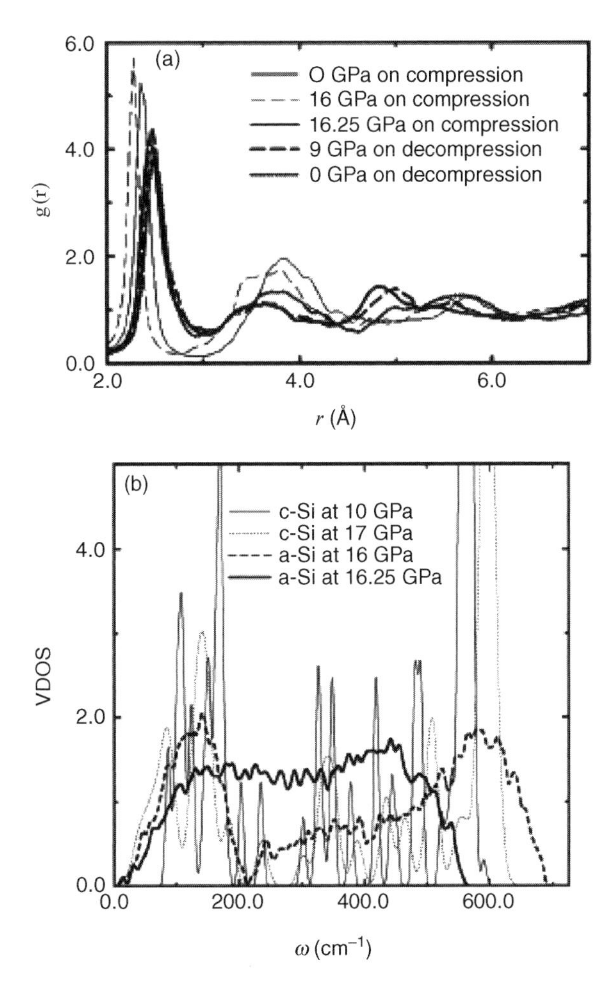


**Figure 18.** (a) Pair correlation functions g(r) for a-Si at various pressures. (b) VDOS for c- and a-Si at various pressures. Both g(r) and VDOS were obtained from the DFT calculations with a local-orbital basis set [266].

has insufficient accuracy to compare it with the theoretical g(r) because of the finite $Q$ range. Partial recrystallization to the β-tin phase during the measurements also made it difficult to refine the raw $S(Q)$ data. These problems prevent us from obtaining real-space structural information such as g(r) and $N_c$. To gain deeper insights into the HDA structure, Daisenberger et al. also performed classic MD simulations of the LDA–HDA transition [267, 268] using the Stillinger–Weber (SW) potential [269]. Their MD results are consistent with their experiment and the previous ab initio MD result [265], and concluded that fivefold-coordinated Si atoms are a major component of the HDA structure.

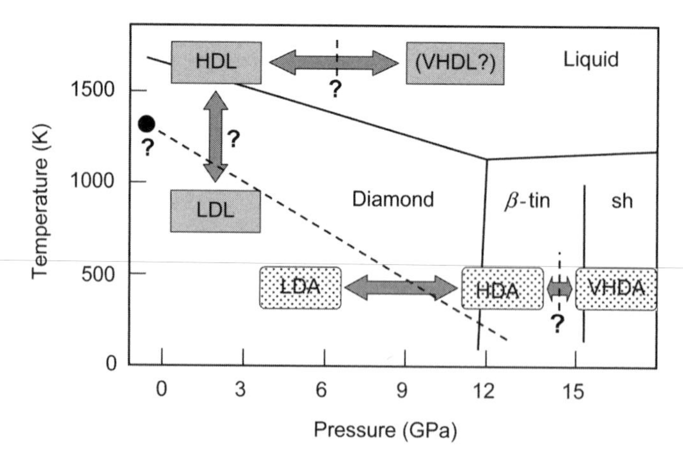


**Figure 19.** Schematic phase relations of Si. Solid lines are the boundaries between liquid and crystalline or crystalline and crystalline phases. Dashed lines denote possible boundaries between amorphous and amorphous or liquid and liquid (metastable) phases. The filled circle denotes the hypothesized second critical point. Note that the scale of pressure and temperature is uncertain.

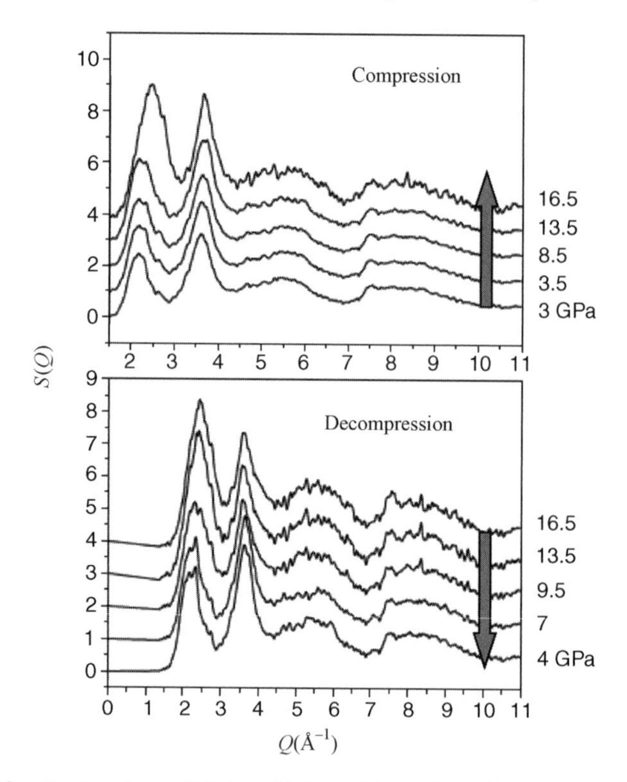


**Figure 20:** Structure factor $S(Q)$ for a-Si obtained from X-ray diffraction experiments [267]. The sample was first compressed up to ~17 GPa and was then decompressed. The LDA–HDA transition took place between 13.5 and 16.5 GPa in the compression process.

Experimental difficulties often lie in obtaining structural data of metastable (amorphous) forms under pressure. Simulation results that complement experimental results are therefore of significant help to interpret experimental data, as demonstrated by Daisenberger et al.

### C. Amorphous Ge

In contrast to what is known about *a*-Si, much less is understood about polyamorphism in Ge. The authors of most early experiments reported no direct evidence of LDA–HDA transition in Ge [260–262, 270, 271]. Shimomura et al. [260] observed a stepwise drop of the electronic resistance (at 6 and 10 GPa) after compression of an *a*-Ge film. This decrease, however, may have resulted from (partial) recrystallization to a metallic high-pressure polymorph under pressure. Tanaka [270] measured X-ray diffraction patterns and optical absorption spectra of *a*-Ge at pressures up to 10 GPa. In this experiment, the sample was indeed partly transformed to the β-tin crystalline phase (~25% in volume) at 6 GPa. Imai et al. [262] also observed an amorphous to β-tin crystal transition. Freund et al. [271], in contrast, have observed no sign of crystallization or transition to an HDA form after compression up to ~9 GPa.

Although LDA–HDA transitions were not observed directly in these experiments, recent experimental [272] and theoretical studies [273] have demonstrated that Ge actually exhibits amorphous–amorphous transition. Using X-ray absorption spectroscopy (XAS), Principi et al. [272] have detected an abrupt change in the local structure and electronic states at ~8 GPa in a compression–decompression cycle of *a*-Ge. The first-neighbor average distance gradually shrank during compression, but it abruptly expanded at ~8 GPa. This structural transition is, in contrast to *a*-Si, irreversible: The modified XAS spectra above 8 GPa was preserved after decompression to 0 GPa (Fig. 21). The obtained new amorphous form shows signatures of a metallic character, and its $N_c$ is roughly estimated to be ~4.5 (increase by ~12%).

Ab initio calculations by Durandurdu and Drabold [273] also support the existence of amorphous–amorphous transition in Ge. They found that LDA Ge was abruptly transformed to an HDA form at 12.75 GPa by using the same computational protocol as that used in their LDA–VHDA transition study of Si [266]. The density of the resultant HDA form is about 20% higher than that of LDA Ge, and its $N_c$ is calculated as ~8. This HDA form therefore strongly resembles the VHDA form of Si, rather than the HDA form of Si. This simulation in passing shows that the transition is irreversible, which is consistent with the experiment by Principi et al. [272].

A tight-binding (TB) MD study to examine the pressure effect on structural and dynamical properties of *a*-Ge has also been reported [274]. The calculations were performed based on the order-*N* nonorthogonal TB framework using the Fermi operator expansion method [275]. The TB MD calculations were run with

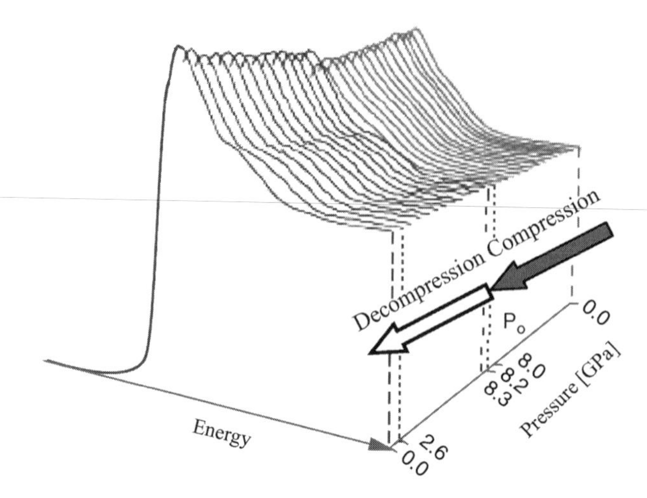


**Figure 21.** X-ray absorption spectroscopy data of a-Ge in a compression–decompression cycle [272]. See color insert.

linear 1% increases in density from 4.79 (LDA Ge) to 7.69 g/cm$^3$. Only a gradual structural change was observed with increasing density, but the first-peak position of g(r) showed anomalous density dependence: It shifted to smaller $r$ as the density was increased up to 6.0 g/cm$^3$ but then began to shift to larger $r$ with the density up to 7.0 g/cm$^3$. Although no abrupt structural change was observed, an HDA form structurally based on the β-tin structure was found to be formed at the density above ~6.5 g/cm$^3$. The β-tin structure of Ge is experimentally stable in a very wide pressure range (10–75 GPa) [255]. Thus, the formation of such a "β-tin-like" HDA form is highly plausible. It is worth noting that, in a recent experiment, vitrification of *l*-Ge under pressure has been attempted, where the "β-tin-like" HDA Ge is expected to form [276].

Unfortunately, the experimental and theoretical studies described above do not provide a consistent picture of polyamorphic transformations of Ge. The ab initio calculations [273] have predicted a highly densified *a*-Ge with $N_c$ of ~8, which may be categorized as VHDA (a third amorphous form). However, considering the stability of the β-tin structure in a wide pressure range, a "β-tin-like" HDA form is more likely to be formed in Ge, as suggested in Refs. [272, 274, and 276]. Additional experimental evidence is eminently desirable.

### D. Liquid Si and Ge (l-Si and l-Ge)

The discovery of multiple amorphous phases has put the spotlight on underlying liquid–liquid (L–L) transitions. Here, attempts thus far to discover possible L–L transitions of Si and Ge will be briefly reviewed. By analogy with L–L transition in water, supercooling *l*-Si or Ge is a promising route to induce possible L–L (HDL–LDL) transitions (see Fig. 19).

Many experiments have been conducted to measure S(*Q*) and density, as well as other thermodynamic properties, such as heat capacity, of supercooled *l*-Si using the containerless levitation technique [277–286]. Almost the same profile of *S*(*Q*) is obtained in all these experiments around the melting temperature (1687 K). However, different tendencies of structural changes during supercooling have been reported among these experiments. For instance, some experiments show that $N_c$ decreases during cooling [277, 279], whereas others show that $N_c$ increases or is retained at a constant down to ~1400 K (Fig. 22) [278, 280, 281]. This inconsistency in structural information may partially come from the enormous difficulty in measuring supercooled *l*-Si in situ, which results in the strong dependence on each experimental measurement. Interestingly, based on their experimental finding that $N_c$ is constant during supercooling, Kim et al. [280] have concluded that Si exhibits no L–L transition.

A recent ab initio MD study [287] has revealed an anomalous structural change of *l*-Si after cooling and has resolved the controversial experimental findings described above. This ab initio (plane wave DFT) study has shown that $N_c$ is indeed constant after supercooling down to ~1400 K, as found experimentally by Kim et al. [280] and Higuchi et al. [281]. However, the calculations have also shown that $N_c$ begins to decrease below ~1200 K, the temperature at which the density maximum occurs [287] (Fig. 22). The tetrahedral network structure is found to grow rapidly during cooling below ~1200 K, which has called Kim's conclusion (no L–L transition) into question (note that no experimental measurement is currently accessible below ~1350 K).

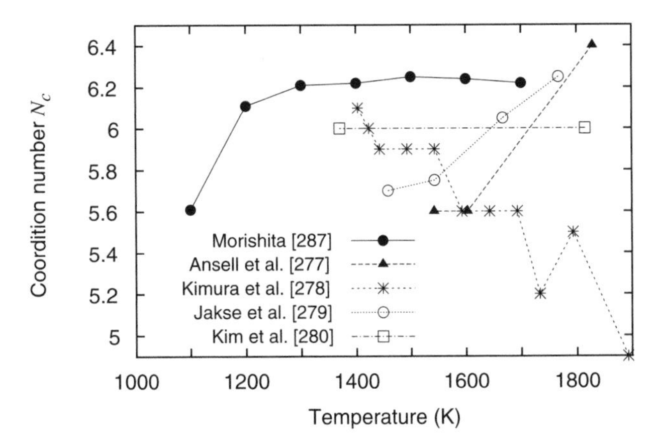


**Figure 22.** Temperature dependence of $N_c$ in *l*-Si during supercooling. All data are experimentally obtained [277–280], except those indicated by the filled circles, which are the ab initio MD results [287].

Clear evidence of L–L transitions has been found only in *l*-Si modeled by the SW potential [269]. Sastry and Angell [288] performed MD simulations of supercooled *l*-Si using the SW potential. After cooling at ambient pressure, the liquid (HDL) was transformed to LDL at ~1060 K. The $N_c$ in LDL is almost 4, and the diffusivity is low compared with that in HDL. The structural properties of LDL, such as g(r) and $N_c$, are very close to those of LDA, which indicates that this HDL–LDL transition is a manifestation of the multiple amorphous forms (LDA and HDA) of Si. McMillan et al. [264] and Morishita [289] have also found structural fluctuations between LDL-like and HDL-like forms in their MD calculations for *l*-Si at 1100 K. Morishita has demonstrated that such a structural fluctuation induces spatial and temporal dynamical heterogeneity, and this heterogeneity accounts for the non-Debye relaxation process that becomes noticeable in the supercooled state [289].

However, we point out that the nature of the L–L transition in SW MD calculations strongly depends on the external conditions and details of the parameters used in the SW potential [290]. For instance, a slight change of the external pressure simply results in a gradual structural change during supercooling [290]. In fact, the SW model gives much lower density than the experimental density for *l*-Si at ambient pressure [289]. An external pressure of 5–6 GPa is necessary to obtain the experimental density in the SW model, but the L–L transition can never be observed under this pressure, and the liquid simply vitrifies to an HDA-like solid [287, 289, 291]. Additional improvements in both experiments and theoretical calculations are thus necessary to clarify whether *l*-Si actually exhibits L–L transitions.

The situation is the same in *l*-Ge. Although X-ray absorption measurements [292] and TB MD calculations [293] have been carried out for supercooled *l*-Ge, L–L transitions have not yet been firmly confirmed.

Pressurization is another possible way to observe L–L transitions or related phenomena (Fig. 19), but many difficulties lie in measuring *l*-Si and *l*-Ge in situ under high pressure. Therefore, only a few experimental studies have been reported to date [294–296]. Tsuji and colleagues measured *l*-Si and *l*-Ge in relatively wide pressure ranges: 4-23 GPa for *l*-Si [294] and 1- 25 GPa for *l*-Ge [295, 296]. Although no discontinuous structural change was observed in either liquid, anomalous pressure dependence of the nearest neighbor distance was revealed in *l*-Si, which expanded about 1.6% between 8 and 14 GPa despite the pressurization. This finding may reflect an underlying drastic structural change at lower temperatures (likely below the melting lines; see Fig. 19). It was also found that anisotropy of the local structure in *l*-Ge, which is estimated by the ratio of the second peak position to the first peak position of $S(Q)$, persists up to 25 GPa.

Simulation studies [297, 298] have supported the experimental results of Tsuji et al. [295]. In addition to structural information, ab initio MD calculations have predicted dynamical properties of *l*-Si under pressure, such as the intermediate scattering function $F(q,t)$ [297] and self-diffusion coefficients $D$ [298]. What should be stressed here is that deeply supercooled *l*-Si exhibits anomalous pressure dependence of the diffusivity [298]. At 1100 K, $D$ increases as the pressure increases up to ~10 GPa, but it begins to decrease above ~10 GPa. This diffusive anomaly has already been observed in water [299, 300] and silicate melts [301–303], and it is attributed to the collapse of the open tetrahedral network by densification. The fact that the diffusive anomaly is also exhibited in *l*-Si strongly suggests that tetrahedrally coordinated liquids (e.g., water, silicate melts, and *l*-Si) share common characteristics that would play a fundamental role in polyamorphism in these substances.

## V. DISCUSSION AND CONCLUSIONS

Overall, recent experimental and theoretical studies have supported the existence of multiple amorphous phases in oxide glasses, water, and semiconductors. Although many parallels exist regarding the amorphous–amorphous transitions, there are also substantial differences. Parallels include the change of coordination number and intermediate range order during compression while retaining an amorphous character. The most prominent difference is the sharp transition with a possible first-order nature from LDA-water to HDA-water, which has not been observed in any other substance. Of course, also temperature and pressure ranges where the transitions are observed differ substantially because of the different interactions involved (e.g., hydrogen-bond interaction versus van der Waals interaction). However, the continuous HDA–VHDA transition in water resembles the broader transitions also observed in silica, even though the nature of intermolecular forces is different.

In the case of oxides, one can conclude that transformations in the glasses in many respects are similar to the transitions that occur in their crystalline counterparts. The most prominent phenomenon in the oxide glasses under compression is coordination increase: 4-to-6 in the case of a-$SiO_2$ and a-$GeO_2$, and 3-to-4 in the case of a-$B_2O_3$. Computer simulations predict an additional increase of coordination number at pressures exceeding 100 GPa—up to 8 for a-$SiO_2$ and a-$GeO_2$, and up to 6 for a-$B_2O_3$, with the analogy of corresponding crystalline substances. These transformations still have to be discovered experimentally in the future. However, several aspects of the phase transition delineate the behavior of oxide glasses from the behavior in the crystals. The coordination transformation occurs in broad pressure intervals and intermediate coordination—five-fold, which does not exist in the crystalline high-pressure modifications, is very important in the case of a-$SiO_2$ and a-$GeO_2$. Second, in all glasses at lower pressures there are transformations, which modify the intermediate-range order (ring-statistics etc.). These transformations are strongly promoted by heating under pressure, and they have no direct analogs in the corresponding crystals.

In the case of water, the three amorphous states LDA, HDA, and VHDA can be distinguished by the number zero, one, or two interstitial water molecules near the faces of the tetrahedron called the "Walrafen pentamer." The tetrahedron itself is not much distorted except for a slight increase of OO-distance caused by the repulsion of interstitial water molecules. Recent studies have suggested that a clear distinction between structurally unrelaxed HDA (uHDA) and structurally relaxed HDA (eHDA or VHDA) has to be made in the future. Although currently it is not entirely clear whether the amorphous ices are glasses structurally related to deeply supercooled liquids or nanocrystals, this distinction may help in sorting out the issue. It is clear now that there is at most one first-order-like transition in amorphous ices and in addition a continuous transition taking place in a finite pressure interval.

In case of Si, it seems that the degree of tetrahedrality is the key to characterizing disordered phases. Degradation of tetrahedrality in HDA Si is moderate, whereas that in VHDA Si is considerable. The striking similarity between the pressure–induced transition sequence of *a*-Si (LDA-HDA-VHDA) and that of *c*-Si (D-β-tin-sh) may give a hint to the essential mechanism for polyamorphism in Si. We would like to stress that many metastable crystalline phases such as BC8 (Si-III) have been observed in Si [252], which indicates that the free energy landscape is likely to be highly complicated and that many variations of each amorphous form may be formed.

Although supporting evidence of the multiple amorphous phases in Ge is not currently sufficient, it seems likely that Ge exhibits amorphous–amorphous and L–L transitions by analogy with Si. It is worth noting that the β-tin phase in Ge is stable in an extremely wide pressure range (10–75 GPa) compared with the β-tin Si phase (10–16 GPa) [252]. The third amorphous phase in Ge (e.g., VHDA Ge) could therefore be expected to form at relatively high pressures (if such a phase exists). It is known that the large 3*d*-electron core radius, which is absent in Si, accounts for the high stability of β-tin Ge over a wide pressure range. Such an effect of the electronic properties on structural stability is of course significant in amorphous solids. Because the interatomic interaction (electronic structure) plays a fundamental role in polyamorphism, a microscopic description that includes the electronic structure is indispensable for profound understanding of polyamorphism, particularly in semiconductors such as Si and Ge.

## Acknowledgments

T.L. is grateful to his M.Sc. and Ph.D. students, Katrin Winkel, Marion Bauer, Michael Elsaesser, Markus Seidl, and Juergen Bernard, for their devotion to the subject and for discussion, to Erwin Mayer for inspiring many experiments and improving the manuscript, to the Austrian Science Fund (FWF projects R37-N11 and Y391), and to the European Research Council (ERC Starting Grant,

SULIWA) for financial support. V.B. is grateful to A.G. Lyapin, O.B. Tsiok, K. Trachenko, and Y. Katayama for their help and useful discussions and to the Russian Foundation for Basic Research (07-02-01275 and 08-02-00014) and the Programs of the Presidium of RAS for financial support. T.M. is grateful to O. Mishima and T. Ikeshoji for useful discussion and to the Ministry of Education, Culture, Sports, Science and Technology, Japan for financial support.

## References

1. R. Zallen, *The Physics of Amorphous Solids*, Wiley, New York, 1983.
2. S. R. Elliott, *Physics of Amorphous Materials*, Longman Group Limited, New York, 1983.
3. J. Zarzycki, *Glasses and the Vitreous State*, Cambridge University Press, Cambridge, UK, 1991.
4. P. M. Ossi, *Disordered Materials—An Introduction*, 2nd ed., Springer-Verlag, Berlin, Germany, 2006.
5. G. P. Johari, *J. Chem. Edu.* **51**, 23 (1974).
6. W. Kauzmann, *Chem. Rev.* **43**, 219 (1948).
7. A. Eisenberg, in Physical Properties of Polymers A. E. James E. Mark, William W. Graessley, Leo Mandelkern, Edward T. Samulski, Jack L. Koenig, and George D. Wignall, eds., American Chemical Society, Washington, DC, 1993.
8. C. A. Angell, K. L. Ngai, G. B. McKenna, P. F. McMillan, and S. W. Martin, *Appl. Phys. Rev.* **88**, 3113 (2000).
9. A. Navrotsky, *Nat. Mater.* **2**, 571 (2003).
10. P. F. McMillan, J. Gryko, C. Bull, R. Arledge, A. J. Kenyon, and B. A. Cressey, *J. Solid State Chem.* **178**, 937 (2005).
11. A. Hedoux, Y. Guinet, and M. Descamps, *Phys. Rev. B: Condens. Matt. Mater. Phys.* **58**, 31 (1998).
12. G. P. Johari, S. Ram, G. Astl, and E. Mayer, *J. Non-Cryst. Solids* **116**, 282 (1990).
13. V. A. Likhachev, *Glass Physics and Chemistry (Translation of Fizika i Khimiya Stekla)* **22**, 80 (1996).
14. S. Susman, K. J. Volin, D. L. Price, M. Grimsditch, J. P. Rino, R. K. Kalia, P. Vashishta, G. Gwanmesia, Y. Wang, and R. C. Liebermann, *Phys. Rev. B* **43**, 1194 (1991).
15. T. Grande, S. Stolen, A. Grzechnik, W. A. Crichton, and M. Mezouar, *Physica A* **314**, 560 (2002).
16. S. Sampath, C. J. Benmore, K. M. Lantzky, J. Neuefeind, K. Leinenweber, D. L. Price, and J. L. Yarger, *Phys. Rev. Lett.* **90**, 115502/1 (2003).
17. Y. Inamura, Y. Katayama, W. Utsumi, and K.-i. Funakoshi, *Phys. Rev. Lett.* **93**, 015501 (2004).
18. J. Urquidi, C. J. Benmore, P. A. Egelstaff, M. Guthrie, S. E. McLain, C. A. Tulk, D. D. Klug, and J. F. C. Turner, *Molec. Phys.* **102**, 2007 (2004).
19. C. J. Benmore, R. T. Hart, Q. Mei, D. L. Price, J. Yarger, C. A. Tulk, and D. D. Klug, *Phys. Rev. B* **72**, 132201 (2005).
20. D. T. Bowron, J. L. Finney, A. Hallbrucker, I. Kohl, T. Loerting, E. Mayer, and A. K. Soper, *J. Chem. Phys.* **125**, (2006).
21. V. V. Brazhkin, *J. Phys. Condens. Matt.* **18**, 9643 (2006).
22. V. V. Brazhkin, *Phys. Rev. Lett.* **98**, 069601 (2007).
23. J. Bernstein, *Polymorphism in Molecular Crystals*, Oxford University Press, Oxford, UK, 2002.
24. V. V. Brazhkin and A. G. Lyapin, *Jetp. Lett-Engl. Tr.* **78**, 542 (2003).
25. O. Mishima, L. D. Calvert, and E. Whalley, *Nature* **314**, 76 (1985).

26. K. Trachenko, V. V. Brazhkin, O. B. Tsiok, M. T. Dove, and E. K. H. Salje, *Phys. Rev. Lett.* **98**, 135502 (2007).
27. H. W. Sheng, H. Z. Liu, Y. Q. Cheng, J. Wen, P. L. Lee, W. K. Luo, S. D. Shastri, and E. Ma, *Nat. Mater.* **6**, 192 (2007).
28. M. C. Wilding, P. F. McMillan, and A. Navrotsky, *Physica A* **314**, 379 (2002).
29. V. V. Brazhkin, Y. Katayama, M. V. Kondrin, T. Hattori, A. G. Lyapin, and H. Saitoh, *Phys. Rev. Lett.* **100**, 145701/1 (2008).
30. S. V. Buldyrev and H. E. Stanley, *Physica A* **330**, 124 (2003).
31. I. Brovchenko, A. Geiger, and A. Oleinikova, *J. Chem. Phys.* **118**, 9473 (2003).
32. I. Brovchenko, A. Geiger, and A. Oleinikova, *J. Chem. Phys.* **123**, 044515 (2005).
33. I. Brovchenko and A. Oleinikova, *J. Chem. Phys.* **124**, 164505 (2006).
34. P. Jedlovszky and R. Vallauri, *J. Chem. Phys.* **122**, 081101/1 (2005).
35. P. Jedlovszky, L. B. Partay, A. P. Bartok, G. Garberoglio, and R. Vallauri, *J. Chem. Phys.* **126**, 241103/1 (2007).
36. P. Jedlovszky, L. B. Partay, A. P. Bartok, V. P. Voloshin, N. N. Medvedev, G. Garberoglio, and R. Vallauri, *J. Chem. Phys.* **128**, 244503/1 (2008).
37. S. M. Sharma and S. K. Sikka, *Prog. Mater. Sci.* **40**, 1 (1996).
38. H. E. Stanley, *Pramana* **53**, 53 (1999).
39. V. V. Brazhkin, A. G. Lyapin, S. V. Popova, and R. N. Voloshin, *NATO Sci. Ser., II Math., Phys. Chem.* **81**, 15 (2002).
40. P. G. Debenedetti, *J. Phys. Condens. Matter* **15**, R1669 (2003).
41. V. V. Brazhkin, and A. G. Lyapin, *J. Phys. Cond. Matter* **15**, 6059 (2003).
42. P. F. McMillan, *J. Mater. Chem.* **14**, 1506 (2004).
43. J. L. Yarger and G. H. Wolf, *Science* **306**, 820 (2004).
44. C. A. Angell, *Annu. Rev. Phys. Chem.* **55**, 559 (2004).
45. S. Petit and G. Coquerel, *Polymorphism* **259** (2006).
46. T. Loerting, K. Winkel, C. G. Salzmann, and E. Mayer, *Phys. Chem. Chem. Phys.* **8**, 2810 (2006).
47. T. Loerting and N. Giovambattista, *J. Phys. Cond. Matt.* **18**, R919 (2006).
48. G. P. Johari and O. Andersson, *Thermochim. Acta* **461**, 14 (2007).
49. W. B. Hubbard, *Planetary Interiors*, Van Nostrand Reinhold, New York, 1984.
50. D. L. Anderson, *Theory of the Earth.*, Blackwell Scientific Publications, Boston, MA, 1989.
51. R. J. Hemley, J. Badro, and D. M. Teter, *Phys. Meets Min.* 173 (2000).
52. A. Putnis, *Introduction to Mineral Sciences*, Cambridge University Press, Cambridge, U.K., 1992.
53. R. J. Hemley, C. T. Prewitt, and K. J. Kingma, *Rev. Mineral.* **29**, 41 (1994).
54. V. Swamy, S. K. Saxena, B. Sundman, and J. Zhang, *J. Geophys. Res.* **99**, 11 (1994).
55. G. A. Lyzenga, T. J. Ahrens, and A. C. Mitchell, *J. Geophys. Res.* **88**, 2431 (1983).
56. D. R. Schmitt and T. J. Ahrens, *J. Geophys. Res.* **94**, 5851 (1989).
57. J. A. Akins and T. J. Ahrens, *Geophys. Res. Lett.* **29**, 31/1 (2002).
58. D. Andrault, G. Fiquet, F. Guyot, and M. Hanfland, *Science* **282**, 720 (1998).
59. S. Ono, *Abstracts of AGU 2001 Fall Meeting, San Francisco*, 2001.

60. L. S. Dubrovinsky, S. K. Saxena, P. Lazor, R. Ahuja, O. Eriksson, J. M. Wills, and B. Johansson, *Nature* **388**, 362 (1997).
61. Y. Kuwayama, K. Hirose, N. Sata, and Y. Ohishi, *Science* **309**, 923 (2005).
62. S. R. Shieh, T. S. Duffy, and G. Shen, *Earth Planet. Sci. Lett.* **235**, 273 (2005).
63. V. P. Prakapenka, G. Shen, L. S. Dubrovinsky, M. L. Rivers, and S. R. Sutton, *J. Phys. Chem. Solids* **65**, 1537 (2004).
64. A. G. Lyapin, V. V. Brazhkin, E. L. Gromnitskaya, O. V. Stal'gorova, and O. B. Tsiok, *Phys. Uspekhi* **42**, 1059 (1999).
65. A. G. Lyapin, V. V. Brazhkin, E. L. Gromnitskaya, V. V. Mukhamadiarov, O. V. Stal'gorova, and O. B. Tsiok, *NATO Sci. Seri., II Math., Phys. Chem.* **81**, 449 (2002).
66. P. W. Bridgman and I. Simon, *J. Appl. Phys.* **24**, 405 (1953).
67. H. M. Cohen and R. Roy, *Phys. Chem. Glasses* **6**, 149 (1965).
68. M. Grimsditch, *Phys. Rev. Lett.* **52**, 2379 (1984).
69. R. J. Hemley, H. K. Mao, P. M. Bell, and B. O. Mysen, *Phys. Rev. Lett.* **57**, 747 (1986).
70. M. Grimsditch, *Phys. Rev. B* **34**, 4372 (1986).
71. T. Gerber, B. Himmel, H. Lorenz, and D. Stachel, *Cryst. Res. Technol.* **23**, 1293 (1988).
72. C. Meade, R. J. Hemley, and H. K. Mao, *Phys. Rev. Lett.* **69**, 1387 (1992).
73. C.-s. Zha, R. J. Hemley, H.-k. Mao, T. D. Duffy, and C. Meade, *Phys. Rev. B: Condens. Matt.* **50**, 13105 (1994).
74. E. M. Stolper and T. J. Ahrens, *Geophys. Res. Lett.* **14**, 1231 (1987).
75. V. G. Karpov and M. Grimsditch, *Phys. Rev. B Condens. Matt.* **48**, 6941 (1993).
76. O. B. Tsiok, V. V. Brazhkin, A. G. Lyapin, and L. G. Khvostantsev, *Phys. Rev. Lett.* **80**, 999 (1998).
77. L. Stixrude and M. S. T. Bukowinski, *Phys. Rev. B Condens. Matt.* **44**, 2523 (1991).
78. J. S. Tse, D. D. Klug, and Y. Le Page, *Phys. Rev. B Condens. Matt.* **46**, 5933 (1992).
79. R. G. Della Valle and E. Venuti, *Phys. Rev. B Condens. Matt.* **54**, 3809 (1996).
80. D. J. Lacks, *Phys. Rev. Lett.* **80**, 5385 (1998).
81. D. J. Lacks, *Phys. Rev. Lett.* **84**, 4629 (2000).
82. E. Demiralp, T. Cagin, and W. A. Goddard, III, *Phys. Rev. Lett.* **82**, 1708 (1999).
83. J. R. Rustad, D. A. Yuen, and F. J. Spera, *Phys. Rev. B Condens. Matt.* **44**, 2108 (1991).
84. Y. Inamura, M. Arai, N. Kitamura, S. M. Bennington, and A. C. Hannon, *Physica B* **241-243**, 903 (1998).
85. G. D. Mukherjee, S. N. Vaidya, and V. Sugandhi, *Phys. Rev. Lett.* **87**, 195501 (2001).
86. P. McMillan, B. Piriou, and R. Couty, *J. Chem. Phys.* **81**, 4234 (1984).
87. T. Rouxel, H. Ji, T. Hammouda, and A. Moreac, *Phys. Rev. Lett.* **100**, 225501/1 (2008).
88. R. A. B. Devine, R. Dupree, I. Farnan, and J. J. Capponi, *Phys. Rev. B Condens. Matt.* **35**, 2560 (1987).
89. V. P. Prakapenka, G. Shen, M. L. Rivers, and S. R. Sutton, *AGU Fall Meet. Suppl. Abstract* **84**, V41E (2003).
90. F. S. El'kin, V. V. Brazhkin, L. G. Khvostantsev, O. B. Tsiok, and A. G. Lyapin, *JETP Lett.* **75**, 342 (2002).
91. K. Trachenko, M. T. Dove, V. Brazhkin, and F. S. El'kin, *Phys. Rev. Lett.* **93**, 135502 (2004).
92. K. Trachenko and M. T. Dove, *Phys. Rev. B Condens. Matt. Mater. Phys.* **67**, 064107/1 (2003).

93. M. F. Thorpe, *J. Non-Cryst. Solids* **57**, 355 (1983).
94. J.-F. Lin, H. Fukui, D. Prendergast, T. Okuchi, Y. Q. Cai, N. Hiraoka, C.-S. Yoo, A. Trave, P. Eng, M. Y. Hu, and P. Chow, *Phys. Rev. B Condens. Matter* **75**, 012201/1 (2007).
95. H. Fukui, M. Kanzaki, N. Hiraoka, and Y. Q. Cai, *Phys. Rev. B Condens. Matt.* **78**, 012203/1 (2008).
96. S. K. Lee, P. J. Eng, H.-k. Mao, Y. Meng, and J. Shu, *Phys. Rev. Lett.* **98**, 105502/1 (2007).
97. S. K. Lee, P. J. Eng, H.-k. Mao, Y. Meng, M. Newville, M. Y. Hu, and J. Shu, *Nature Mat.* **4**, 851 (2005).
98. V. V. Brazhkin and K. Trachenko, unpublished data.
99. M. Kanzaki, *J. Am. Ceram. Soc.* **73**, 3706 (1990).
100. E. Ohtani, F. Taulelle, and C. A. Angell, *Nature* **314**, 78 (1985).
101. X. Xue, J. F. Stebbins, M. Kanzaki, and R. G. Tronnes, *Science* **245**, 962 (1989).
102. J. Zhang, R. C. Liebermann, T. Gasparik, C. T. Herzberg, and Y. Fei, *J. Geophys. Res.* **98**, 19785 (1993).
103. I. Saika-Voivod, F. Sciortino, and P. H. Poole, *Phys. Rev. E* **63**, 11202 (2000).
104. J. Horbach, *J. Phys. Condens. Matt.* **20**, 244118/1 (2008).
105. P. K. Hung, N. V. Hong, and L. T. Vinh, *J. Phys. Condens. Matt.* **19**, 466103/1 (2007).
106. V. B. Prakapenka, L. S. Dubrovinsky, G. Shen, M. L. Rivers, S. R. Sutton, V. Dmitriev, H. P. Weber, and T. Le Bihan, *Phys. Rev. B Condens. Matt.* **67**, 132101/1 (2003).
107. S. Ono, T. Tsuchiya, K. Hirose, and Y. Ohishi, *Phys. Rev. B Condens. Matt.* **68**, 014103/1 (2003).
108. C. E. Stone, A. C. Hannon, T. Ishihara, N. Kitamura, Y. Shirakawa, R. N. Sinclair, N. Umesaki, and A. C. Wright, *J. Non-Cryst. Solids* **293-295**, 769 (2001).
109. J. P. Itie, A. Polian, G. Calas, J. Petiau, A. Fontaine, and H. Tolentino, *Phys. Rev. Lett.* **63**, 398 (1989).
110. K. H. Smith, E. Shero, A. Chizmeshya, and G. H. Wolf, *J. Chem. Phys.* **102**, 6851 (1995).
111. M. Guthrie, C. A. Tulk, C. J. Benmore, J. Xu, J. L. Yarger, D. D. Klug, J. S. Tse, H. K. Mao, and R. J. Hemley, *Phys. Rev. Lett.* **93**, 115502/1 (2004).
112. X. Hong, G. Shen, V. B. Prakapenka, M. Newville, M. L. Rivers, and S. R. Sutton, *Phys. Rev. B Condens. Matt.* **75**, 104201/1 (2007).
113. M. Micoulaut, L. Cormier, and G. S. Henderson, *J. Phys. Condens. Matt.* **18**, R753 (2006).
114. K. V. Shanavas, N. Garg, and S. M. Sharma, *Phys. Rev. B Condens. Matt.* **73**, 094120/1 (2006).
115. G. Shen, H.-P. Liermann, S. Sinogeikin, W. Yang, X. Hong, C.-S. Yoo, and H. Cynn, *Proc. Natl. Acad. Sci. U.S.A.* **104**, 14576 (2007).
116. A. G. Lyapin, V. V. Brazhkin, Y. Katayama, and Y. Inamura, *Abstracts of Joint 21st AIRAPT and 45th EHPRG International Conference, Catania*, 2007, 246.
117. O. Ohtaka, H. Arima, H. Fukui, W. Utsumi, Y. Katayama, and A. Yoshiasa, *Phys. Rev. Lett.* **92**, 155506/1 (2004).
118. V. V. Hoang, N. H. T. Anh, and H. Zung, *Phys. B* **390**, 17 (2007).
119. J. Krogh-Moe, *J. Non-Cryst. Solids* **1**, 269 (1969).
120. D. R. Uhlmann, J. F. Hays, and D. Turnbull, *Phys. Chem. Glasses* **8**, 1 (1967).
121. J. D. Mackenzie, *J. Am. Ceram. Soc.* **46**, 461 (1963).
122. A. C. Wright, C. E. Stone, R. N. Sinclair, N. Umesaki, N. Kitamura, K. Ura, N. Ohtori, and A. C. Hannon, *Phys. Chem. Glasses* **41**, 296 (2000).
123. E. Chason and F. Spaepen, *J. Appl. Phys.* **64**, 4435 (1988).

124. M. Grimsditch, R. Bhadra, and Y. Meng, *Phys. Rev. B Condens. Matt. Mater. Phys.* **38**, 7836 (1988).

125. J. Nicholas, S. Sinogeikin, J. Kieffer, and J. Bass, *Phys. Rev. Lett.* **92**, 215701/1 (2004).

126. M. Grimsditch, A. Polian, and A. C. Wright, *Phys. Rev. B Condens. Matt.* **54**, 152 (1996).

127. A. Takada, *Phys. Chem. Glasses* **45**, 156 (2004).

128. L. Huang, J. Nicholas, J. Kieffer, and J. Bass, *J. Phys. Condens. Matter* **20**, 075107/1 (2008).

129. V. V. Brazhkin, Y. Katayama, K. Trachenko, O. B. Tsiok, A. G. Lyapin, E. Artacho, M. Dove, G. Ferlat, Y. Inamura, and H. Saitoh, *Phys. Rev. Lett.* **101**, 035702 (2008).

130. T. Hattori, T. Kinoshita, T. Narushima, K. Tsuji, and Y. Katayama, *Phys. Rev. B Condens. Matt. Mater. Phys.* **73**, 054203/1 (2006).

131. G. Widmann and R. Riesen, *J. Thermal Anal. Calorim.* **52**, 109 (1998).

132. E. F. Burton and W. F. Oliver, *Proc. R. Soc. Ser. A* **153**, 166 (1935).

133. E. F. Burton and W. F. Oliver, *Nature* **135**, 505 (1935).

134. E. Mayer and R. Pletzer, *Nature* **319**, 298 (1986).

135. R. Pletzer and E. Mayer, *J. Chem. Phys.* **90**, 5207 (1989).

136. E. Mayer and R. Pletzer, *J. Phys. Colloq.* **48**, 581 (1987).

137. Z. Dohnalek, G. A. Kimmel, P. Ayotte, R. S. Smith, and B. D. Kay, *J. Chem. Phys.* **118**, 364 (2003).

138. A. Hallbrucker, E. Mayer, and G. P. Johari, *J. Phys. Chem.* **93**, 4986 (1989).

139. P. Jenniskens and D. F. Blake, *Science* **265**, 753 (1994).

140. P. Jenniskens, D. F. Blake, M. A. Wilson, and A. Pohorille, *Astrophys. J.* **455**, 389 (1995).

141. P. Jenniskens and D. F. Blake, *Planet. Space Sci.* **44**, 711 (1996).

142. B. Guillot and Y. Guissani, *J. Chem. Phys.* **120**, 4366 (2004).

143. P. Parent, S. Lacombe, F. Bournel, and C. Laffon, *Physics and Chemistry of Ice*, W. F. Kuhs, ed., RSC Publishing, Cambridge, U.K., 2007.

144. P. Brggeller and E. Mayer, *Nature* **288**, 569 (1980).

145. E. Mayer and P. Brueggeller, *Nature* **298**, 715 (1982).

146. E. Mayer, *J. Appl. Phys.*, **58**, 663 (1985).

147. V. F. Petrenko and R. W. Whitworth, *Physics of Ice*, Oxford University Press, Oxford, U.K., 1999.

148. M. Bauer, M. S. Elsaesser, K. Winkel, E. Mayer, and T. Loerting, *Phys. Rev. B* **77**, 220105/1 (2008).

149. O. Mishima, L. D. Calvert, and E. Whalley, *Nature* **310**, 393 (1984).

150. T. Loerting, I. Kohl, W. Schustereder, A. Hallbrucker, and E. Mayer, *Chem. Phys. Chem.* **7**, 1203 (2006).

151. J. S. Tse, D. D. Klug, C. A. Tulk, I. Swainson, E. C. Svensson, C. K. Loong, V. Shpakov, V. R. Belosludov, R. V. Belosludov, and Y. Kawazoe, *Nature* **400**, 647 (1999).

152. T. Loerting, C. Salzmann, I. Kohl, E. Mayer, and A. Hallbrucker, *Phys. Chem. Chem. Phys.* **3**, 5355 (2001).

153. O. Mishima, *J. Chem. Phys.* **100**, 5910 (1994).

154. K. Winkel, M. S. Elsaesser, M. Seidl, M. Bauer, E. Mayer, and T. Loerting, *J. Phys. Cond. Matt.* in press.

155. K. Winkel, M. S. Elsaesser, E. Mayer, and T. Loerting, *J. Chem. Phys.* **128**, 044510/1 (2008).

156. C. G. Salzmann, E. Mayer, and A. Hallbrucker, *Phys. Chem. Chem. Phys.* **6**, 1269 (2004).

157. J. E. Bertie, L. D. Calvert, and E. Whalley, *J. Chem. Phys.* **38**, 840 (1963).
158. J. E. Bertie, L. D. Calvert, and E. Whalley, *Can. J. Chem.* **42**, 1373 (1964).
159. D. D. Klug, Y. P. Handa, J. S. Tse, and E. Whalley, *J. Chem. Phys.* **90**, 2390 (1989).
160. A. I. Kolesnikov, V. V. Sinitsyn, E. G. Ponyatovsky, I. Natkaniec, L. S. Smirnov, and J. C. Li, *J. Phys. Chem. B* **101**, 6082 (1997).
161. Y. Yoshimura, S. T. Stewart, M. Somayazulu, H.-k. Mao, and R. J. Hemley, *J. Chem. Phys.* **124**, 024502 (2006).
162. C. McBride, C. Vega, E. Sanz, and J. L. F. Abascal, *J. Chem. Phys.* **121**, 11907 (2004).
163. Y. Yoshimura, H.-k. Mao, and R. J. Hemley, *Chem. Phys. Lett.* **420**, 503 (2006).
164. N. Sartori, J. Bednar, and J. Dubochet, *J. Microscop.* **182**, 163 (1996).
165. G. A. Baratta, G. Leto, F. Spinella, G. Strazzulla, and G. Foti, *Astron. Astrophys.* **252**, 421 (1991).
166. A. Kouchi and T. Kuroda, *Nature* **344**, 134 (1990).
167. G. Leto and G. A. Baratta, *Astron. Astrophys.* **397**, 7 (2003).
168. M. H. Moore and R. L. Hudson, *Astrophys. J.* **401**, 353 (1992).
169. R. L. Hudson and M. H. Moore, *J. Phys. Chem.* **96**, 6500 (1992).
170. O. Mishima, *Nature* **384**, 546 (1996).
171. J. L. Finney, D. T. Bowron, A. K. Soper, T. Loerting, E. Mayer, and A. Hallbrucker, *Phys. Rev. Lett.* **89**, 205503 (2002).
172. C. G. Salzmann, T. Loerting, S. Klotz, P. W. Mirwald, A. Hallbrucker, and E. Mayer, *Phys. Chem. Chem. Phys.* **8**, 386 (2006).
173. T. Loerting, W. Schustereder, K. Winkel, C. G. Salzmann, I. Kohl, and E. Mayer, *Phys. Rev. Lett.* **96**, 025702 (2006).
174. K. Winkel, W. Schustereder, I. Kohl, C. G. Salzmann, E. Mayer, and T. Loerting, *Proc. 11th Intl. Conf. on the Physics and Chemistry of Ice*, W. F. Kuhs, ed., RSC, Dorchester, U.K., 2007, pp. 641.
175. M. C. Bellissent-Funel, L. Bosio, A. Hallbrucker, E. Mayer, and R. Sridi-Dorbez, *J. Chem. Phys.* **97**, 1282 (1992).
176. D. D. Klug, C. A. Tulk, E. C. Svensson, and C. K. Loong, *Phys. Rev. Lett.* **83**, 2584 (1999).
177. J. L. Finney, A. Hallbrucker, I. Kohl, A. K. Soper, and D. T. Bowron, *Phys. Rev. Lett.* **88**, 225503 (2002).
178. J. A. Ripmeester, C. I. Ratcliffe, and D. D. Klug, *J. Chem. Phys.* **96**, 8503 (1992).
179. D. D. Klug, O. Mishima, and E. Whalley, *J. Chem. Phys.* **86**, 5323 (1987).
180. M. C. Bellissent-Funel, and L. Bosio, *J. Chem. Phys.* **102**, 3727 (1995).
181. M. C. Bellissent-Funel, *Europhys. Lett.* **42**, 161 (1998).
182. F. W. Starr, M.-C. Bellissent-Funel, and H. E. Stanley, *Phys. Rev. E* **60**, 1084 (1999).
183. M. C. Bellissent-Funel, J. Teixeira, and L. Bosio, *J. Chem. Phys.* **87**, 2231 (1987).
184. E. G. Ponyatovsky, V. V. Sinitsyn, and T. A. Pozdnyakova, *J. Chem. Phys.* **109**, 2413 (1998).
185. M. Sasai and E. Shiratani, *Nippon Kessho Gakkaishi* **40**, 101 (1998).
186. V. V. Sinitsyn, E. G. Ponyatovsky, A. I. Kolesnikov, U. Dahlborg, and M. Calvo-Dahlborg, *Solid State Ionics* **145**, 415 (2001).
187. E. L. Gromnitskaya, O. V. Stal'gorova, V. V. Brazhkin, and A. G. Lyapin, *Phys. Rev. B* **64**, 094205 (2001).

188. C. A. Tulk, C. J. Benmore, J. Urquidi, D. D. Klug, J. Neuefeind, B. Tomberli, and P. A. Egelstaff, *Science* **297**, 1320 (2002).
189. M. M. Koza, H. Schober, H. E. Fischer, T. Hansen, and F. Fujara, *J. Phys. Condens. Matt.* **15**, 321 (2003).
190. K. Winkel, E. Mayer, and T. Loerting, in press.
191. M. M. Koza, B. Geil, K. Winkel, C. Koehler, F. Czeschka, M. Scheuermann, H. Schober, and T. Hansen, *Phys. Rev. Lett.* **94**, 125506 (2005).
192. M. M. Koza, T. Hansen, R. P. May, and H. Schober, *J. Non-Cryst. Solids* **352**, 4988 (2006).
193. M. Scheuermann, B. Geil, K. Winkel, and F. Fujara, *J. Chem. Phys.* **124**, 224503/1 (2006).
194. O. Mishima and Y. Suzuki, *Nature* **419**, 599 (2002).
195. S. Klotz, T. Straessle, R. J. Nelmes, J. S. Loveday, G. Hamel, G. Rousse, B. Canny, J. C. Chervin, and A. M. Saitta, *Phys. Rev. Lett.* **94**, 025506 (2005).
196. O. Mishima and H. E. Stanley, *Nature* **396**, 329 (1998).
197. C. A. Tulk, C. J. Benmore, D. D. Klug, and J. Neuefeind, *Phys. Rev. Lett.* **96**, 149601 (2006).
198. C. G. Salzmann, E. Mayer, and A. Hallbrucker, *Phys. Chem. Chem. Phys.* **6**, 5156 (2004).
199. R. J. Nelmes, J. S. Loveday, T. Straessle, C. L. Bull, M. Guthrie, G. Hamel, and S. Klotz, *Nat. Phys.* **2**, 414 (2006).
200. T. Loerting, K. Winkel, and E. Mayer, in press.
201. G. P. Johari, *J. Chem. Phys.* **121**, 8428 (2004).
202. S. Sastry, P. G. Debenedetti, F. Sciortino, and H. E. Stanley, *Phys. Rev. E* **53**, 6144 (1996).
203. P. G. Debenedetti, *Nature* **392**, 127 (1998).
204. R. J. Hemley, L. C. Chen, and H. K. Mao, *Nature* **338**, 638 (1989).
205. O. Andersson, *Phys Rev. Lett.* **95**, 205503 (2005).
206. O. Andersson and A. Inaba, *Phys. Rev. B* **74** (2006).
207. R. J. Speedy, *J. Phys. Chem.* **86**, 3002 (1982).
208. R. J. Speedy, *J. Phys. Chem.* **86**, 982 (1982).
209. P. H. Poole, F. Sciortino, U. Essmann, and H. E. Stanley, *Nature* **360**, 324 (1992).
210. H. E. Stanley, C. A. Angell, U. Essmann, M. Hemmati, P. H. Poole, and F. Sciortino, *Physica A* **206**, 1 (1994).
211. L. P. N. Rebelo, P. G. Debenedetti, and S. Sastry, *J. Chem. Phys.* **109**, 626 (1998).
212. C. A. Angell, *Science* **319**, 582 (2008).
213. J. S. Tse, D. M. Shaw, D. D. Klug, S. Patchkovskii, G. Vankó, G. Monaco, and M. Krisch, *Phys. Rev. Lett.* **100**, 095502 (2008).
214. J. C. Li and P. Jenniskens, *Planet. and Space Sci.* **45**, 469 (1997).
215. G. P. Johari, *J. Chem. Phys.* **112**, 8573 (2000).
216. B. Geil, M. M. Koza, F. Fujara, H. Schober, and F. Natali, *Phys. Chem. Chem. Phys.* **6**, 677 (2004).
217. H. Schober, M. M. Koza, A. Tolle, C. Masciovecchio, F. Sette, and F. Fujara, *Phys. Rev. Lett.* **85**, 4100 (2000).
218. O. Andersson and H. Suga, *Phys Rev. B* **65**, 140201 (2002).
219. C. G. Salzmann, I. Kohl, T. Loerting, E. Mayer, and A. Hallbrucker, *Phys. Chem. Chem. Phys.* **5**, 3507 (2003).
220. A. Hallbrucker, E. Mayer, and G. P. Johari, *J. Phys. Chem.* **93**, 7751 (1989).
221. G. P. Johari, A. Hallbrucker, and E. Mayer, *J. Phys. Chem.* **94**, 1212 (1990).

222. E. Mayer, in *J. Mol. Struct.*, **250**, 403 (1991).
223. G. P. Johari, A. Hallbrucker, and E. Mayer, *Nature* **330**, 552 (1987).
224. A. Hallbrucker, E. Mayer, and G. P. Johari, *Philos. Mag. B*, **60**, 179 (1989).
225. I. Kohl, L. Bachmann, E. Mayer, A. Hallbrucker, and T. Loerting, *Nature* **435**, E1 (2005).
226. I. Kohl, L. Bachmann, A. Hallbrucker, E. Mayer, and T. Loerting, *Phys. Chem. Chem. Phys.* **7**, 3210 (2005).
227. M. Chonde, M. Brindza, and V. Sadtchenko, *J. Chem. Phys.* **125**, 094501/1 (2006).
228. G. P. Johari, *J. Chem. Phys.* **127**, 157101 (2007).
229. M. Fisher and J. P. Devlin, *J. Phys. Chem.*, **99**, 11584 (1995).
230. R. S. Smith and B. D. Kay, *Nature* **398**, 788 (1999).
231. R. S. Smith, Z. Dohnalek, G. A. Kimmel, K. P. Stevenson, and B. D. Kay, *Chem. Phys.*, **258**, 291 (2000).
232. S. M. McClure, E. T. Barlow, M. C. Akin, D. J. Safarik, T. M. Truskett, and C. B. Mullins, *J. Phys. Chem. B* **110**, 17987 (2006).
233. S. M. McClure, D. J. Safarik, T. M. Truskett, and C. B. Mullins, *J. Phys. Chem. B* **110**, 11033 (2006).
234. P. Jenniskens and D. F. Blake, *Astrophy. J.* **473**, 1104 (1996).
235. P. Jenniskens, S. F. Banham, D. F. Blake, and M. R. S. McCoustra, *J. Chem. Phys.* **107**, 1232 (1997).
236. N. Giovambattista, C. A. Angell, F. Sciortino, and H. E. Stanley, *Phys Rev. Lett.* **93**, 047801/1 (2004).
237. Y. Yue and C. A. Angell, *Nature* **427**, 717 (2004).
238. C. A. Angell, *J. Phys. Condens. Matt.* **19**, 205112/1 (2007).
239. G. P. Johari, *Phys. Chem. Chem. Phys.* **2**, 1567 (2000).
240. D. D. Klug, *Nature* **420**, 749 (2002).
241. J. S. Tse and M. L. Klein, *Phys. Rev. Lett.* **58**, 1672 (1987).
242. A. I. Kolesnikov, V. V. Sinitsyn, E. G. Ponyatovsky, I. Natkaniec, and L. S. Smirnov, *Phys. B Condens. Matt.* **213-214**, 474 (1995).
243. H. Schober, M. Koza, A. Tolle, F. Fujara, C. A. Angell, and R. Bohmer, *Phys. B Condens. Matt.* **241-243**, 897 (1998).
244. C. A. Tulk, D. D. Klug, E. C. Svensson, V. F. Sears, and J. Katsaras, *Appl. Phys. A Mater. Sci. Proc.* **74**, S1185 (2002).
245. O. Andersson and A. Inaba, *J. Chem. Phys.* **122**, 124710/1 (2005).
246. N. I. Agladze and A. J. Sievers, *Phys. B Condens. Matt.* **316-317**, 513 (2002).
247. T. Strassle, A. M. Saitta, S. Klotz, and M. Braden, **93** (2004).
248. E. L. Gromnitskaya, O. V. Stal'gorova, A. G. Lyapin, V. V. Brazhkin, and O. B. Tarutin, *JETP Lett.* **78**, 488 (2003).
249. T. Strassle, S. Klotz, G. Hamel, M. M. Koza, and H. Schober, *Phys. Rev. Lett.* **99**, 175501/1 (2007).
250. G. P. Johari and O. Andersson, *Phys. Rev. B Condens. Matt. Mater. Phys.* **70**, 184108/1 (2004).
251. O. Andersson, *J. Phys. Cond. Matt.* **20**, 244115 (2008).
252. O. Mishima, *J. Chem. Phys.* **115**, 4199 (2001).
253. O. Mishima, *J. Chem. Phys.* **121**, 3161 (2004).
254. H. E. Stanley, S. V. Buldyrev, N. Giovambattista, E. La Nave, A. Scala, F. Sciortino, and F. W. Starr, in *New Kinds of Phase Transitions: Transformations in Disordered Substances, Vol.*

*NATO Science Series, Vol. 81,* V. V. Brazhkin, S. V. Buldyrev, V. N. Ryzhov, and H. E. Stanley, eds., Kluwer Academic Publishers, Dordrecht, Germany, 2002, p. 309.

255. A. Mujica, A. Rubio, A. Munoz, and R. J. Needs, *Rev. Mod. Phys.* **75**, 863 (2003).
256. E. Rapoport, *J. Chem. Phys.* **46**, 2891 (1967).
257. E. G. Ponyatovskii and O. I. Barkalov, *Mat. Sci. Rep.* **8**, 147 (1992).
258. C. A. Angell and C. T. Moynihan, *Metallurg. Mat. Transact. B* **31B**, 587 (2000).
259. C. T. Moynihan and C. A. Angell, *J. Non-Cryst. Solids* **274**, 131 (2000).
260. O. Shimomura, S. Minomura, N. Sakai, K. Asaumi, K. Tamura, J. Fukushima, and H. Endo, *Phil. Mag.* **29**, 547 (1974).
261. O. Shimomura, in *High Pressure and Low Temperature Physics,* C. W. Chu and J. A. Woollam, eds., Plenum Press, New York, 1978, p. 483.
262. M. Imai, T. Mitamura, K. Yaoita, and K. Tsuji, *High Press. Res.* **15**, 167 (1996).
263. S. K. Deb, M. Wilding, M. Somayazulu, and P. F. McMillan, *Nature* **414**, 528 (2001).
264. P. F. McMillan, M. Wilson, D. Daisenberger, and D. Machon, *Nat. Mat.* **4**, 680 (2005).
265. T. Morishita, *Phys. Rev. Lett.* **93**, 055503 (2004).
266. M. Durandurdu and D. A. Drabold, *Phys. Rev. B Condens. Matt.* **64**, 014101 (2001).
267. D. Daisenberger, M. Wilson, P. F. McMillan, R. Q. Cabrera, M. C. Wilding, and D. Machon, *Phys. Rev. B* **75**, 224118 (2007).
268. P. F. McMillan, M. Wilson, M. C. Wilding, D. Daisenberger, M. Mezouar, and G. N. Greaves, *J. Phys. Condens. Matter* **19**, 415101/1 (2007).
269. F. H. Stillinger and T. A. Weber, *Phys. Rev. B Condens. Matt.* **31**, 5262 (1985).
270. K. Tanaka, *Phys. Rev. B Condens. Matt.* **43**, 4302 (1991).
271. J. Freund, R. Ingalls, and E. D. Crozier, *J. Phys. Chem.* **94**, 1087 (1990).
272. E. Principi, A. Di Cicco, F. Decremps, A. Polian, S. De Panfilis, and A. Filipponi, *Phys. Rev. B* **69**, 201201 (2004).
273. M. Durandurdu and D. A. Drabold, *Phys. Rev. B* **66**, 041201 (2002).
274. J. Koga, K. Nishio, T. Yamaguchi, and F. Yonezawa, *J. Phys. Soc. Jpn.* **73**, 388 (2004).
275. S. Goedecker and L. Colombo, *Phys. Rev. Lett.* **73**, 122 (1994).
276. M. H. Bhat, V. Molinero, E. Soignard, V. C. Solomon, S. Sastry, J. L. Yarger, and C. A. Angell, *Nature* **448**, 787 (2007).
277. S. Ansell, S. Krishnan, J. J. Felten, and D. L. Price, *J. Phys. Condens. Matt.* **10**, L73 (1998).
278. H. Kimura, M. Watanabe, K. Izumi, T. Hibiya, D. Holland-Moritz, T. Schenk, K. R. Bauchspiess, S. Schneider, I. Egry, K. Funakoshi, and M. Hanfland, *Appl. Phys. Lett.* **78**, 604 (2001).
279. N. Jakse, L. Hennet, D. L. Price, S. Krishnan, T. Key, E. Artacho, B. Glorieux, A. Pasturel, and M.-L. Saboungi, *Appl. Phys. Lett.* **83**, 4734 (2003).
280. T. H. Kim, G. W. Lee, B. Sieve, A. K. Gangopadhyay, R. W. Hyers, T. J. Rathz, J. R. Rogers, D. S. Robinson, K. F. Kelton, and A. I. Goldman, *Phys. Rev. Lett.* **95**, 085501/1 (2005).
281. K. Higuchi, K. Kimura, A. Mizuno, M. Watanabe, Y. Katayama, and K. Kuribayashi, *Meas. Sci. Technol.* **16**, 381 (2005).
282. K. Ohsaka, S. K. Chung, W. K. Rhim, and J. C. Holzer, *Appl. Phys. Lett.* **70**, 423 (1997).
283. W. K. Rhim, S. K. Chung, A. J. Rulison, and R. E. Spjut, *Int. J. Thermophys.* **18**, 459 (1997).
284. M. Langen, T. Hibiya, M. Eguchi, and I. Egry, *J. Cryst. Growth* **186**, 550 (1998).
285. W. K. W. K. Rhim and K. Ohsaka, *J. Cryst. Growth* **208**, 313 (2000).

286. Z. Zhou, S. Mukherjee, and W.-K. Rhim, *J. Cryst. Growth* **257**, 350 (2003).
287. T. Morishita, *Phys. Rev. Lett.* **97**, 165502/1 (2006).
288. S. Sastry and C. A. Angell, *Nat. Mat.* **2**, 739 (2003).
289. T. Morishita, *Phys. Rev. E* **77**, 020501/1 (2008).
290. P. Beaucage and N. Mousseau, *J. Phys. Condens. Matt.* **17**, 2269 (2005).
291. T. Morishita, unpublished.
292. A. Filipponi and A. Di Cicco, *Phys. Rev. B* **51**, 12322 (1995).
293. J. Koga, K. Nishio, T. Yamaguchi, and F. Yonezawa, *J. Phys. Soc. Jpn.* **73**, 136 (2004).
294. N. Funamori and K. Tsuji, *Phys. Rev. Lett.* **88**, 255508 (2002).
295. K. Tsuji, T. Hattori, T. Mori, T. Kinoshita, T. Narushima, and N. Funamori, *J. Phys. Condens. Matter* **16**, S989 (2004).
296. M. Ohtani, K. Tsuji, H. Nosaka, N. Hosokawa, and N. Funamori, *Proc. 17th AIRAPT Conference*, Hawaii 1999, p. 498.
297. A. Delisle, D. J. Gonzalez, and M. J. Stott, *J. Phys. Condens. Matt.* **18**, 3591 (2006).
298. T. Morishita, *Phys. Rev. E* **72**, 021201 (2005).
299. C. A. Angell, E. D. Finch, L. A. Woolf, and P. Bach, *J. Chem. Phys.* **65**, 3063 (1976).
300. F. X. Prielmeier, E. W. Lang, R. J. Speedy, and H. D. Ludemann, *Phys. Rev. Lett.* **59**, 1128 (1987).
301. I. Kushiro, *J. Geophys. Res.* **81**, 6347 (1976).
302. I. Kushiro and H. S. Yoder, Jr., B. O. Mysen, *J. Geophys. Res.* **81**, 6351 (1976).
303. G. H. Wolf and P. F. McMillan, in *Reviews in Mineralogy,* J. F. Stebbins, P. F. McMillan, and D. B. Dingwell, eds., Mineralogical Society of America, Washington, DC, 1995, p. 505.

# HYDRATION DYNAMICS AND COUPLED WATER–PROTEIN FLUCTUATIONS PROBED BY INTRINSIC TRYPTOPHAN

DONGPING ZHONG

*Departments of Physics, Chemistry, and Biochemistry, The Ohio State University Columbus, Ohio 43210*

## CONTENTS



*Advances in Chemical Physics, Volume 143*, edited by Stuart A. Rice

## I. INTRODUCTION

Water is a natural solvent and the lubricant of life. Almost all proteins are inactive in the absence of water [1, 2]. The hydration of a protein is particularly important for structural stability and for biological function [1–12]. Understanding the structure and dynamics of protein-associated water molecules (biological water) is fundamental to biology. In the hydration layer of a protein, water molecules are very heterogeneous and have different spatial and temporal characteristics that are defined by the landscape of the protein and by the kinetics of processes [9–17]. A range of experimental techniques has been used over the past decades to tackle this complex problem. Early dielectric measurements of protein solutions indicated the existence of a hydration layer around protein surfaces and suggested different water dynamics in the layer [18, 19]. X-ray diffraction from single protein crystals by electron density maps revealed interior water molecules buried in the protein and bound water molecules trapped in crevices [20, 21]. These water molecules have typical residence times longer than 1 ns at hydration sites. Neutron diffraction from protein powders also uncovered both bound and dynamic water molecules by the detection of hydrogen atoms [22, 23]. Both diffraction techniques reported beautiful ordered-water structures. Nuclear magnetic resonance (NMR) studies of hydration in protein solution reported the associated-water residence times as well as their structures [21, 24–26]. The observed interior water molecules are consistent with the X-ray observations. Two methods were extensively used, which include the Nuclear Overhauser Effect (NOE) with a time resolution on the sub-nanosecond scale from 300 ps to 500 ps [27], and nuclear magnetic relaxation dispersion (NMRD) with a shorter time resolution, reaching a limit of picoseconds [28]. Although both methods can be used to detect certain mobile water molecules, such studies are not directly time resolved, and the inherent lack of spatial resolution and the complication of hydrogen exchange make the molecular interpretation a subject of discussion. Nevertheless, those NMR studies provided important information about the very fast dynamics of some surface water molecules around proteins.

Molecular dynamics (MD) simulations [29–31], coupled with experimental observations, have played an important role in the understanding of protein hydration. They predicted that the dynamics of ordered water molecules in the surface layer is ultrafast, typically on the picosecond time scales. Most calculated residence times are shorter than experimental measurements reported before, in a range of sub-picosecond to $\sim$100 ps. Water molecules at the surface are very mobile and are in constant exchange with bulk water. For example, the trajectory study of myoglobin hydration revealed that among 294 hydration sites, the residence times at 284 sites (96.6% of surface water molecules) are less than 100 ps [32]. Furthermore, the population time correlation functions obtained for different hydration sites all showed a biphasic distribution, approximately a double-exponential decay. One component is on the time scale of ~1 ps, and the other is in the range of 10–100 ps. The molecular origin of these two distinct time distributions has not been fully understood, and their biological significance is not yet known.

Femtosecond spectroscopy has an ideal temporal resolution for the study of ultrafast water motions from femtosecond to picosecond time scales [33–36]. Femtosecond solvation dynamics is sensitive to both time and length scales and can be a good probe for protein hydration dynamics [16, 37–50]. Recent femtosecond studies by an extrinsic labeling of a protein with a dye molecule showed certain ultrafast water motions [37–42]. This kind of labeling usually relies on hydrophobic interactions, and the probe is typically located in the hydrophobic crevice. The resulting dynamics mostly reflects bound water behavior. The recent success of incorporating a synthetic fluorescent amino acid into the protein showed another way to probe protein electrostatic interactions [43, 48].

We recently developed a systematic method that uses the intrinsic tryptophan residue (Trp or W) as a local optical probe [49, 50]. Using site-directed mutagenesis, tryptophan can be mutated into different positions one at a time to scan protein surfaces. With femtosecond temporal and single-residue spatial resolution, the fluorescence Stokes shift of the local excited Trp can be followed in real time, and thus, the location, dynamics, and functional roles of protein–water interactions can be studied directly. With MD simulations, the solvation by water and protein (residues) is differentiated carefully to determine the hydration dynamics. Here, we focus our own work and review our recent systematic studies on hydration dynamics and protein–water fluctuations in a series of biological systems using the powerful intrinsic tryptophan as a local optical probe, and thus reveal the dynamic role of hydrating water molecules around proteins, which is a longstanding unresolved problem and a topic central to protein science.

## II. OPTICAL PROBE OF INTRINSIC AMINO ACID TRYPTOPHAN

### A. Characterization of Excited-State Tryptophan

#### 1. *Photophysics*

The studies of water motions around proteins have been difficult because of the lack of a reliable optical probe. Early attempts to use an extrinsic labeling of a protein with a dye molecule showed certain mobile water molecules [37–42]. Such extrinsic labeling has limited probing sites and thus prevents general applications. We proposed to use the intrinsic tryptophan residue as a local solvation probe. Using powerful site-directed mutations, tryptophan can be mutated into any desirable places one at a time. Thus, the protein contains only one Trp residue, and we can achieve site-specific probing of local water motions, which is not accessible to any other techniques. Such a strategy provides a versatile method to study water motions and protein dynamics in various biological systems.

Tryptophan has been used extensively to study protein dynamics with its lifetime, emission peak, quantum yield, and rotational anisotropy [51–56]. Among the 20 natural amino acid residues, tryptophan has the longest absorption wavelength, peaking at 280 nm, with a considerable fluorescence quantum yield of 0.14 in water. Because of limited time resolution and technique difficulty, early efforts unfortunately implied that tryptophan was not a practical ultrafast optical probe based on its complex photophysics [57–60]. Until recently, tryptophan was examined carefully again [61, 62], and it is realized now that tryptophan can be a powerful optical probe [49, 50]. In the ground state, tryptophan has a static dipole of 1.87 D [63]; see Fig. 1. The first two excited states, $^1L_b$ ($S_1$) and $^1L_a$ ($S_2$), are nearly degenerate in vacuum and have perpendicular transition moments. The $^1L_a$ state has a larger static dipole, 6.12 D, and the $^1L_b$ state has 1.55 D [63]. Thus, as shown in Fig. 2, the $^1L_a$ state in polar solvent already lies below $^1L_b$ after excitation before environment relaxation. After relaxation, the $^1L_a$ state is much lower than $^1L_b$, and the observed fluorescence of tryptophan is dominantly from the $^1L_a$ emission. The femtosecond- and wavelength-resolved fluorescence transients of tryptophan in water are shown in Fig. 3a. Similar to many other dyes used for probing bulk solvation [33–35], the transients show systematic decays at the blue side of the emission peak and rises at the red side, which is a typical signature of solvation dynamics. The photophysics of tryptophan has recently been characterized carefully, and the internal conversion of $^1L_b$ to $^1L_a$ has been shown to occur

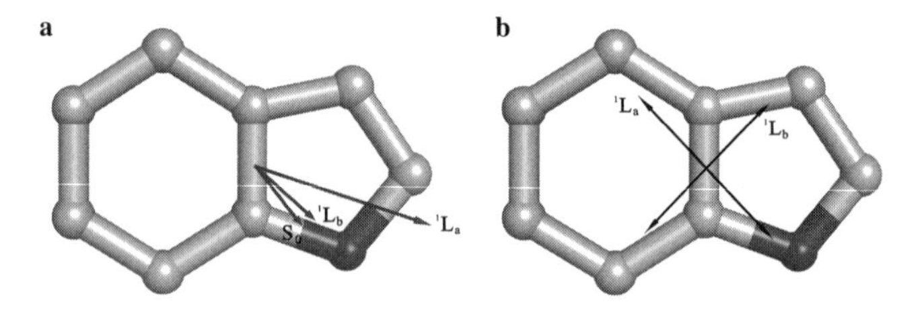


**Figure 1.** Structure of the indole moiety of side-chain tryptophan. White and black balls represent carbon and nitrogen atoms, respectively. The permanent dipole moments of indole in the ground state $S_0$ and two low-lying electronic excited states ($^1L_a$ and $^1L_b$) are shown as single arrows in (a), and the transition dipole moments between ground and excited states are shown as double arrows in (b).

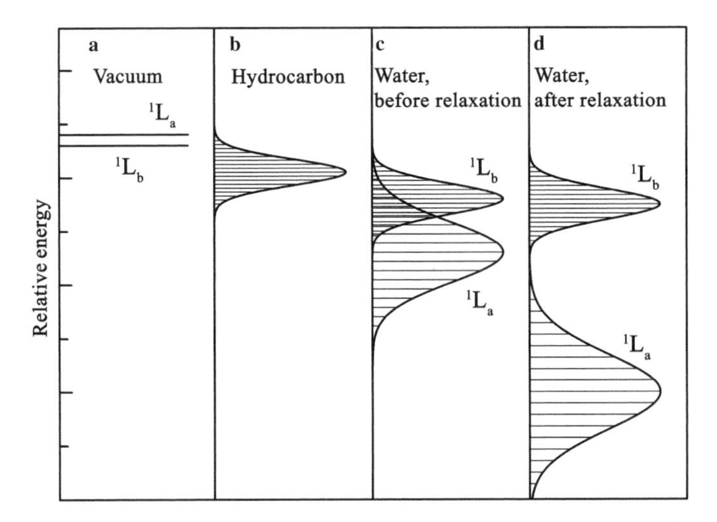


**Figure 2.** Schematic representation of inhomogeneous broadening and relative energy levels of $^1L_a$ and $^1L_b$ states (a) under vacuum, (b) in nonpolar hydrocarbon solvent, (c) in polar water at the instant of absorption, and (d) in polar water after certain-time solvation [55].

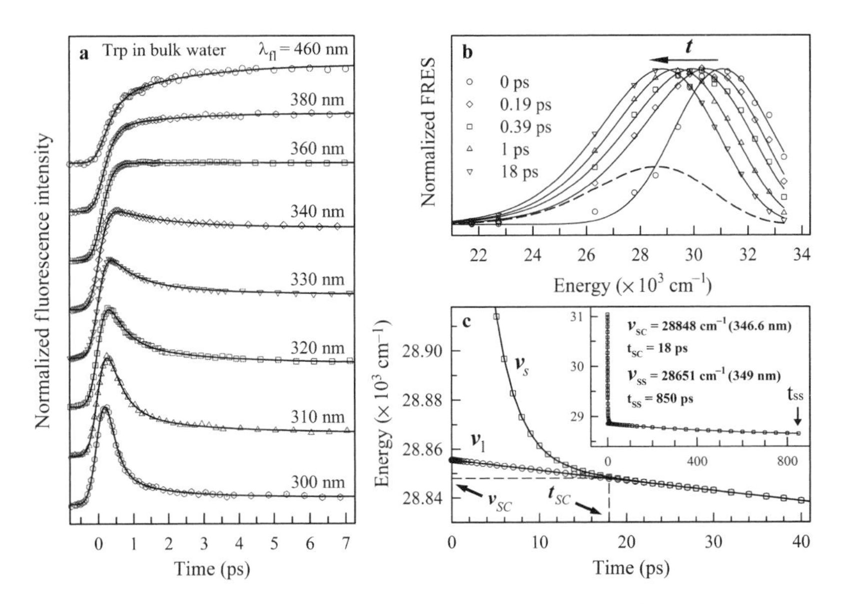


**Figure 3.** (a) Normalized fs-resolved fluorescence transients of tryptophan in bulk water with a series of wavelength detection. The symbols are raw data, and the solid lines are best multiexponential fit. (b) Normalized FRES of tryptophan in bulk water at several representative times. The dashed curve is the steady-state emission spectrum with the peak at 349 nm. (c) The emission maximum changes ($\nu_s$ and $\nu_l$) with time constructed from the overall FRES in (b) and from the two lifetime-associated FRES. The two functions merge at $\nu_{sc}$ at $t_{sc}$, and solvation dynamics is completed. The spectrum keeps evolution to reach the steady-state emission $\nu_{ss}$ at $t_{ss}$.

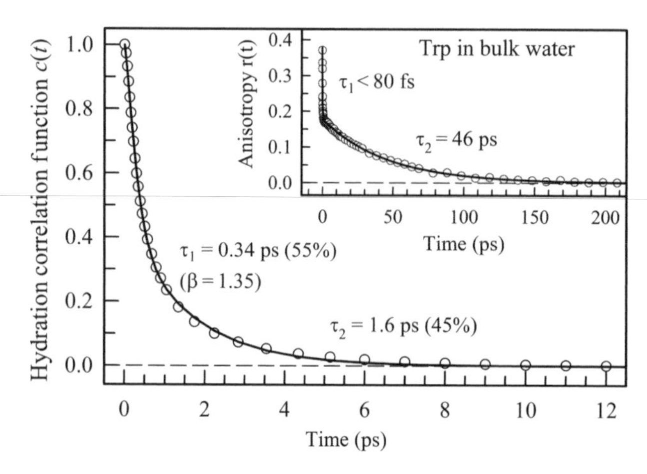


**Figure 4.** Ultrafast hydration correlation function $c(t)$ of tryptophan. The $c(t)$ can be fit with a stretched biexponential model as shown. The inset shows the fluorescence anisotropy dynamics of tryptophan after ultrafast ultraviolet (UV) absorption. The internal conversion between $^1L_a$ and $^1L_b$ states occurs within 80 fs. The free rotation time of tryptophan in bulk water is 46 ps.

ultrafast, in less than 100 fs [61, 62], as also shown by the ultrafast anisotropy decay in the inset of Fig. 4. In the ground state, tryptophan was believed to populate at least two different rotamers. The rotamer dynamics, which is in the range of ~500 ps to several nanoseconds based on proton- and electron-transfer quenching, has also been studied extensively [57]. In fact, such a rotamer model should be considered as simplification of intrinsic dynamic heterogeneity of tryptophan. Nevertheless, the observed excited-state dynamics, i.e., the internal conversion in less than 100 fs and the rotamer dynamics in longer than 500 ps, make tryptophan a powerful optical probe with a wide time window from subpicosecond to subnanosecond, and more importantly, many dynamic processes of proteins occur in this time range.

### 2. *Quenching Residues*

With site-directed mutation and femtosecond-resolved fluorescence methods, we have used tryptophan as an excellent local molecular reporter for studies of a series of ultrafast protein dynamics, which include intraprotein electron transfer [64–68] and energy transfer [61, 69], as well as protein hydration dynamics [70–74]. As an optical probe, all these ultrafast measurements require no potential quenching of excited-state tryptophan by neighboring protein residues or peptide bonds on the picosecond time scale. However, it is known that tryptophan fluorescence is readily quenched by various amino acid residues [75] and peptide bonds [76–78]. Intraprotein electron transfer from excited indole moiety to nearby electrophilic residue(s) was proposed to be the quenching mechanism [75–78]. We have systematically studied various ultrafast quenching dynamics, especially within 100 ps, and identified important quenching residues from our extensive examinations of more than 40 proteins [79]. Each protein only contains a single tryptophan residue by site-specific mutation. Among 20 amino acids, we observed two major ultrafast quenchers: carbonyl- and sulfur-containing groups. The former (-C=O) includes the residues of aspartate, glutamate, asparagine, and glutamine as well as the peptide bond; the latter is the cysteine residue (-SH) and related disulfide bond (-S-S-). Surprisingly, recent extensive analyses of high-resolution protein structures in the Protein Data Bank showed a strong tendency of weak interactions between aromatic residues with carbonyl oxygen atoms [80] and thiol or disulfide sulfur atoms [81, 82]. These close weak contacts at a short distance of ~3–4 Å are believed to stabilize the protein structure and may have a functional role [80–82]. Thus, identification of these ultrafast quenching residues and determination of their time scales are essential to using tryptophan as a local optical probe for ultrafast protein dynamics.

## B. Experimental Methodology

### 1. *Data Analysis*

Solvation dynamics are measured using the more reliable energy relaxation method after a local perturbation [83–85], typically using a femtosecond-resolved fluorescence technique. Experimentally, the wavelength-resolved transients are obtained using the fluorescence upconversion method [85]. The observed fluorescence dynamics, decay at the blue side and rise at the red side (Fig. 3a), reflecting typical solvation processes. The molecular mechanism is schematically shown in Fig. 5. Typically, by following the standard procedures [35], we can construct the femtosecond-resolved emission spectra (FRES, Stokes shifts with time) and then the correlation function (solvent response curve):

$$c(t) = \frac{\nu(t) - \nu(\infty)}{\nu(0) - \nu(\infty)} \tag{1}$$

where $\nu(t)$, $\nu(0)$, and $\nu(\infty)$ are femtosecond-resolved emission maximum in $\text{cm}^{-1}$ at different times, respectively. For a molecular probe with only *one* lifetime, $\nu(\infty)$ usually equals to the steady-state emission maximum $\nu_{ss}$. However, in a simple model, tryptophan in the ground state can be treated phenomenologically to populate at least two rotamers, and these excited rotamers in water emit with two different lifetimes (500 ps and 3 ns) and two different emission maxima ($\nu_1$, $\nu_2$). Thus, $\nu(\infty)$ *cannot* be considered to be the apparent $\nu_{ss}$. We need to find out when solvation is completed ($t_{sc}$) and the corresponding emission maximum ($\nu_{sc}$). In the following discussion, we present

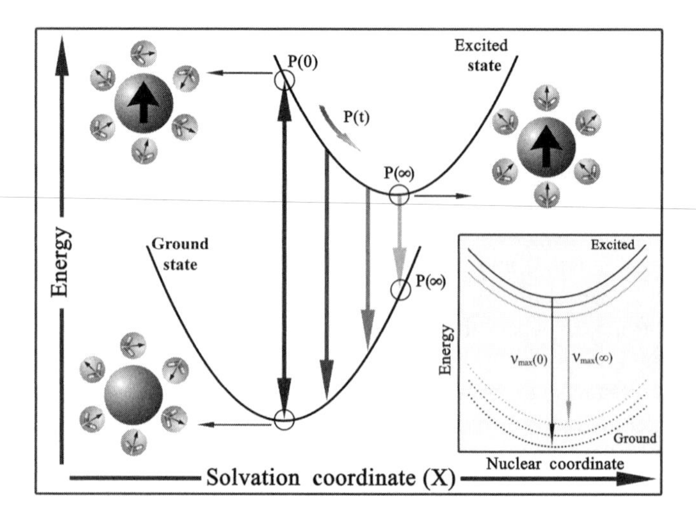


**Figure 5.** Schematic illustration of the potential energy surfaces involved in solvation dynamics showing the water orientational motions along the solvation coordinate together with instantaneous polarization P. In the inset, we show the change in the potential energy along the intramolecular nuclear coordinate. As solvation proceeds, the energy of the solute comes down, which causes a red shift in the fluorescence spectrum [9].

an extended method to determine $t_{sc}$ and $v_{sc}$ from femtosecond-resolved fluorescence measurements [49].

All femtosecond-resolved transients in Fig. 3a can be best fitted by a sum of a series of exponential functions. These functions can be separated into two parts: One part represents solvation processes, and the other one is for lifetime emissions (population decay). The transient signal can be written as follows:

$$I_{\lambda}(t) = I_{\lambda}^{solv}(t) + I_{\lambda}^{popul}(t) = \sum_{i} a_i e^{-t/\tau_i} + \sum_{j} b_j e^{-t/\tau_j} \tag{2}$$

where the first term is for solvation and the second term is for lifetime emission. The coefficient $a_i$ is positive (decay dynamics) at the blue side of the emission peak ($<349$ nm) and is negative (initial rise) at the red side ($\geq 349$ nm). The coefficient $b_j$ is always positive and represents relative contributions of two lifetime emissions (500 ps and 3 ns). The *overall* FRES can be constructed as follows:

$$I(\lambda, t) = \frac{I_{\lambda}^{ss} I_{\lambda}(t)}{\sum_{i} a_i \tau_i + \sum_{j} b_j \tau_j}, \quad I(v, t) = v^{-2} I(\lambda, t) \tag{3}$$

where $I_\lambda^{ss}$ is the steady-state relative emission intensity at $\lambda$. For a given $t$, the emission spectrum can be constructed from the gated transients. The resulting FRES of tryptophan in bulk water are shown in Fig. 3b. These spectra are fitted using a log-norm function to deduce the emission maximum $\nu(t)$. Thus, a function of emission maxima with time ($\nu_s$) can be obtained, which is shown in Fig. 3c. At a certain time ($t_{sc}$), solvation is complete and the emission maximum $\nu_s(t_{sc})$ should be equal to the *apparent* lifetime emission peak $\nu_l(t_{sc})$ (not steady-state peak yet), which results from a mixture of two lifetime emissions and is constructed as follows:

$$I^{popul}(\lambda, t) = \frac{I_\lambda^{ss} I_\lambda^{popul}(t)}{\sum\limits_i a_i \tau_i + \sum\limits_j b_j \tau_j}, \qquad I^{popul}(\nu, t) = \nu^{-2} I^{popul}(\lambda, t) \tag{4}$$

The lifetime-associated FRES $\nu_l$, also shown in Fig. 3c, merges with $\nu_s$ at $t_{sc}$, and the value of $\nu_{sc}$ at the merging point could be taken as $\nu(\infty)$ in Eq. (1). Thus,

$$c(t) = \frac{\nu_s(t) - \nu_{sc}}{\nu_s(0) - \nu_{sc}} \tag{5}$$

where $c(t)$ stops at $t_{sc}$. But a more accurate way is to subtract the apparent lifetime emission maximum $\nu_l(t)$ from the overall emission maximum $\nu_s(t)$ at any given $t$, and the resulting $c(t)$ is written as follows:

$$c(t) = \frac{\nu_s(t) - \nu_l(t)}{\nu_s(0) - \nu_l(0)} \tag{6}$$

When solvation time is much shorter than the lifetimes, both constructions of $c(t)$ give very similar results because $\nu_l(t) \approx \nu_{sc}$. However, when solvation dynamics becomes slower, such as in proteins, on a time scale close to the lifetimes, the contribution of $\nu_l(t)$ is significant, and Eq. (6) must be used to construct $c(t)$. For all results reported here, we used Eq. (6). Note that for the molecular dye probe with only single lifetime emission, $\nu_l(t) = \nu_l(0) = \nu_{ss} = \nu(\infty)$ and Eq. (6) becomes equal to Eq. (1).

The key step here is to construct both $\nu_s(t)$ and $\nu_l(t)$ and to determine $t_{sc}$; see Fig. 3c. When the difference of two maxima $(\nu_s(t) - \nu_l(t))$ reaches 0.5 cm$^{-1}$, we consider solvation complete ($t = t_{sc}$). The difference between $\nu_{sc}$ and $\nu_{ss}$ purely results from the mixture of two lifetime fluorescence emissions. The time evolution from $\nu_{sc}$ to $\nu_{ss}$ could be very long. For tryptophan in water, the time-zero emission maximum ($\nu_0$) is obtained at 322.1 nm and solvation is completed in 18 ps. Both $\nu_s$ and $\nu_l$ merge at 346.6 nm, and the total Stokes shift is 2186 cm$^{-1}$. However, it takes another 832 ps for the emission spectrum to reach the steady-state maximum at 349 nm (only 197 $cm^{-1}$ shift because of the two-emission mixture). In a sense, the steady-state emission maximum is not relevant for construction of the correlation function. The lifetime emission spectrum has a maximum at 329.8 nm for the 500-ps component and at 349.8 nm for the 3-ns component, which is consistent with previous results [57].

The solvation dynamics of bulk water have been well studied. Jarzeba et al. [33] obtained a correlation function with 160 fs (33%) and 1.2 ps (67%), and Jimemez et al. [34] reported an initial Gaussian-type component (frequency 38.5 $ps^{-1}$ $\approx$25 fs in time width, $\sim$48%) and two exponential decays of 126 fs (20%) and 880 fs (35%). Using Eq. (6), the correlation function we obtained for bulk water, as shown in Fig. 4, is best fitted by double exponential decays integrated with an initial Gaussian-type contribution through a stretched mode: $c(t) = c_1 e^{-(t/\tau_1)^\beta} + c_2 e^{-t/\tau_2}$, where for a pure Gaussian-type decay, $\beta = 2$. The final fitting results are $c_1 = 55\%$, $\tau_1 = 340$ fs, $\beta = 1.35$, $c_2 = 45\%$, and $\tau_2 = 1.6$ ps. The 1.6-ps component observed here, which is a little longer than previously reported values ($\sim$1 ps), is probably caused by the stronger interactions between water molecules and Zwitterionic tryptophan (pH = 7.0) in the first solvation shell.

### 2. *Site-Directed Mutations*

The establishment of intrinsic tryptophan as a local optical probe of hydration dynamics is significant. As mentioned previously, with the protein engineering method, we can design mutations specifically into desirable places of interest. Because we have screened potential quenching residues [79], we can avoid these residue quenchers during design. One concern was raised about the local perturbation induced by mutation. The optical probe tryptophan can sense the environment's response in a range of about 10 Å through dipole–dipole/charge interactions [70], which mainly probes the motions of local water networks not only for Trp-associated water molecules. Tryptophan is so far the smallest molecule that could be used as an optical probe in proteins, compared with the often-labeled fluorescent dye coumarin 153 [41], 2-(p-toluidino)naphthalene-6-sulfonate (TNS) and 5-(dimethylamino)naphthalene-1-sulfonyl chloride (DC) [42], or eosin [40] and the synthetic fluorescent amino acid Aladan [43]. More importantly, tryptophan is an intrinsic amino acid, not extrinsic or synthetic, and the labeling can be site specific using site-directed mutagenesis. We carefully choose residues to be mutated, which should have similar properties as to hydrophobicity and size as tryptophan to minimize the perturbation induced by the latter and also should have no hydrogen bonding with neighboring residues to avoid a change of structure and stability. Overall, tryptophan is probably the best optical molecular probe used so far in the study of protein solvation. Thus, the perturbation induced by Trp mutation, although unavoidable, will not significantly affect the hydration dynamics of local hydrogen-bond networks.

# III. MODEL SYSTEMS

In this section, we review our first examinations of tryptophan probing sensitivity and water dynamics in a series of important model systems from simple to complex, which range from a tripeptide [70], to a prototype membrane protein melittin [70], to a common drug transporter human serum albumin [71], and to lipid interface of a nanochannel [86]. At the end, we also give a special case that using indole moiety of tryptophan probes supramolecule crown ether solvation, and we observed solvent-induced supramolecule folding [87]. The obtained solvation dynamics in these systems are linked to properties or functions of these biological-relevant macromolecules.

## A. Tripeptide Lysine-Tryptophan-Lysine

Figure 6 shows a snapshot of the molecular structure of the tripeptide Lysine-Tryptophan-Lysine (KWK) in water with 2-ns MD simulations. Both $\pi$-cation and hydrophobic interactions are clearly present in this folded structure in recognition. The original idea was to examine $\pi$-cation interaction and solvation dynamics around this structural motif. Figure 7 shows fluorescence temporal behaviors of tryptophan in tripeptide-KWK, and the overall decay dynamics are similar to that of tryptophan in the same buffer [49]. Besides the two apparent lifetime contributions, three exponential decays were used to represent the solvation dynamics. At the blue side, the three decays occur in 0.22–0.4, 1.8–3.4,

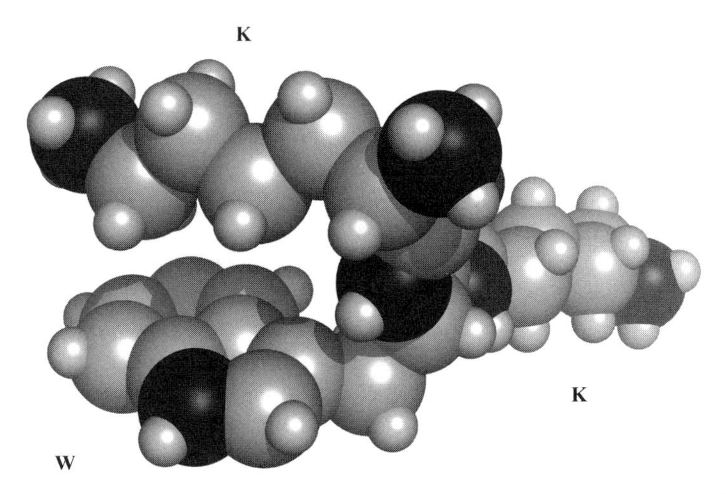


**Figure 6.** A snapshot of tripeptide-KWK structure generated in 2-ns trajectory MD simulations. The small balls are H atoms. The big balls in light gray, dark gray, and black are C, O, and N atoms, respectively.

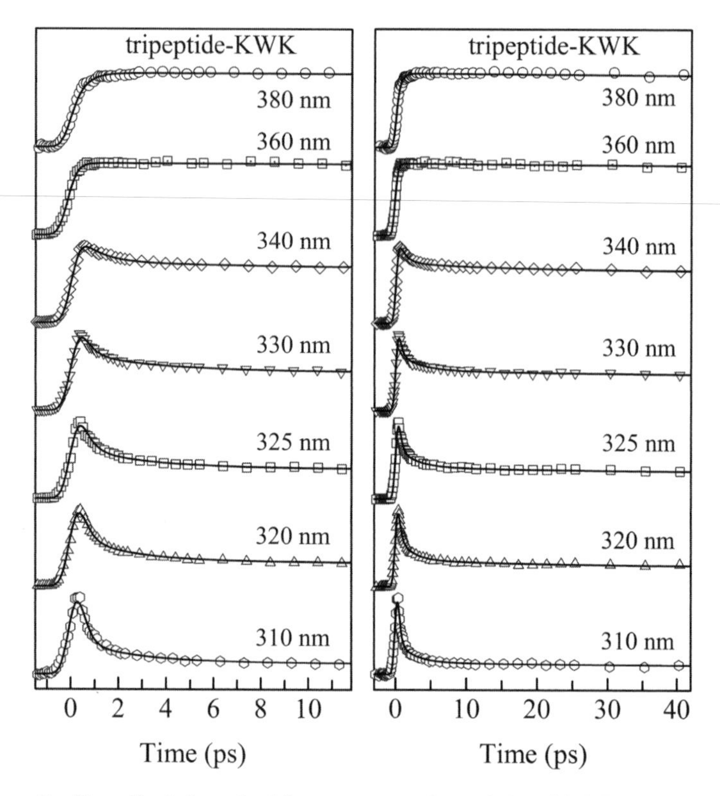


**Figure 7.** Normalized, fs-resolved fluorescence transients of tripeptide-KWK in the short (left) and long (right) time ranges with a series of gated fluorescence emissions.

and 8–24 ps, and the first ultrafast component dominates. At the red side ($\geq$350 nm), an initial rise was observed in all transients within 0.2–0.5 ps.

The solvation correlation function was constructed using Eq. (6), and the derived result is shown in Fig. 8. The solvent response function can be represented by three exponential decays: 0.58 ps with a 66% of total amplitude, 3.1 ps (27%), and 23 ps (7%). Free tryptophan in the same buffer gives a relaxation distribution of 0.52 ps (68%), 1.9 ps (23%), and 7.6 ps (9%) [49]. The first dominant ultrafast component represents bulk-like water motion. The third minor component (23 ps and 7%) probably results from some rigid water molecules sticking to the positively charged lysine in close proximity of tryptophan. Overall, no significant formation of ordered-water layers is observed, and water is mostly bulk like. Thus, the interactions of this tripeptide with DNA and proteins involve minimal desolvation energy, and the recognition would be fast. Interestingly, we did not observe any local wobbling motion of tryptophan in tripeptide-KWK as shown in the inset of Fig. 8. The time scale of 127 ps from the tripeptide represents the

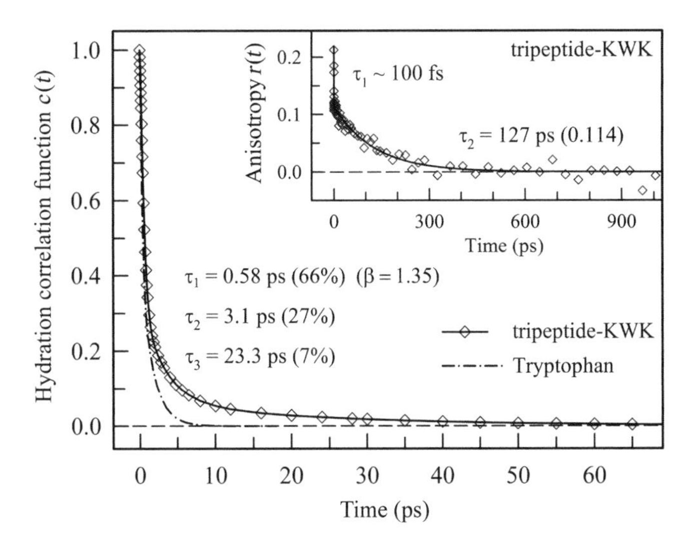


**Figure 8.** Hydration correlation function $c(t)$ of tripeptide-KWK. The $c(t)$ can be fit with a stretched triexponential decay function as shown. For comparison, the hydration correlation function of tryptophan in bulk water is also shown. The inset shows the fluorescence anisotropy dynamics of KWK. The initial ~100-fs dynamics resulted from the internal conversion between $^1L_a$ and $^1L_b$ states, and the free rotation time of KWK in bulk water is 127 ps.

overall tumbling motion of the entire peptide in solution, which indicates a rigid local structure probably caused by tryptophan–lysine(s) cation–$\pi$ interactions [88, 89]. However, our recent MD simulations showed a very dynamic heterogeneous structure of KWK with three isomers in the ground state with a free energy difference of less than 5.2 kJ/mol and suggested that the long response component could be from structural isomerization [90].

## B. A Prototype Membrane Protein Melittin

Melittin, which is an amphipathic peptide from honeybee venom, consists of 26 amino acid residues and adopts different conformations from a random coil, to an $\alpha$-helix, and to a self-assembled tetramer under certain aqueous environments; see Fig. 9. We have carried out our systematic studies of the hydration dynamics in these three conformations using a single intrinsic tryptophan (W19) as a molecular probe. The folded $\alpha$-helix melittin was formed with lipid interactions to mimic physiological membrane-bound conditions. The self-assembled tetramer was prepared under high-salt concentration (NaCl = 2 M). The tryptophan emission of three structures under three different aqueous environments is 348.5 nm, 341 nm, and 333.5 nm, which represents different exposures of aqueous solution from complete in random-coil, to locating at the lipid surface of a nanochannel (50 Å in diameter) in $\alpha$-helix and to partially buried in tetramer. Figure 10 shows

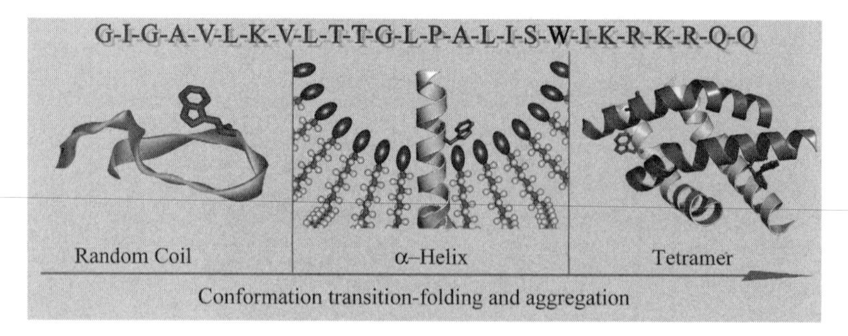


**Figure 9.** Structures of melittin in three different conformations: random coil, membrane-bound $\alpha$-helix, and tetramer. A sequence of melittin is shown on top, and tryptophans in three conformations are shown as dark sticks.

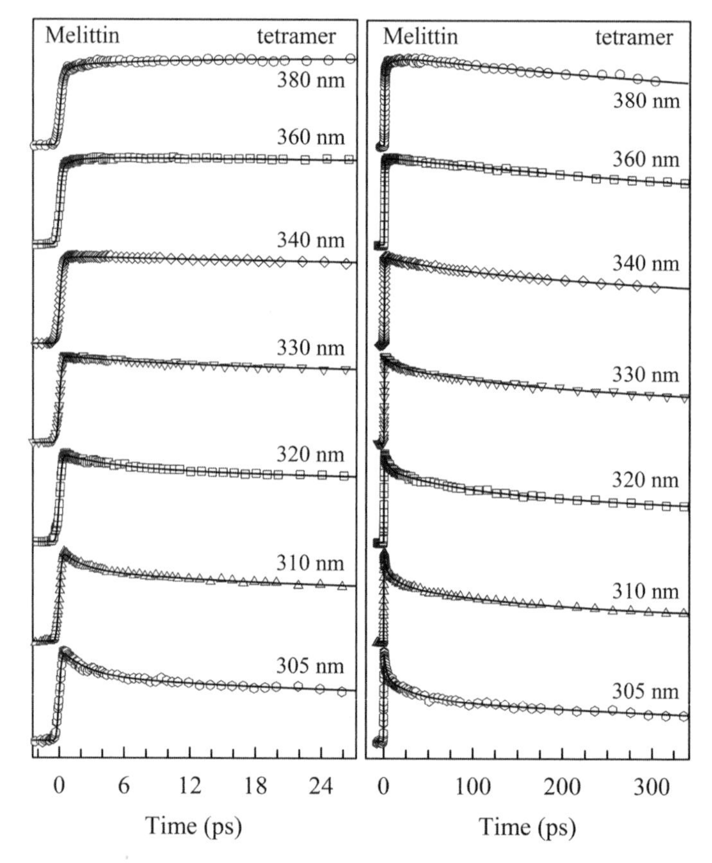


**Figure 10.** Normalized, fs-resolved fluorescence transients of a melittin tetramer formed in high-salt aqueous solution in the short (left) and long (right) time ranges with a series of gated fluorescence emissions.

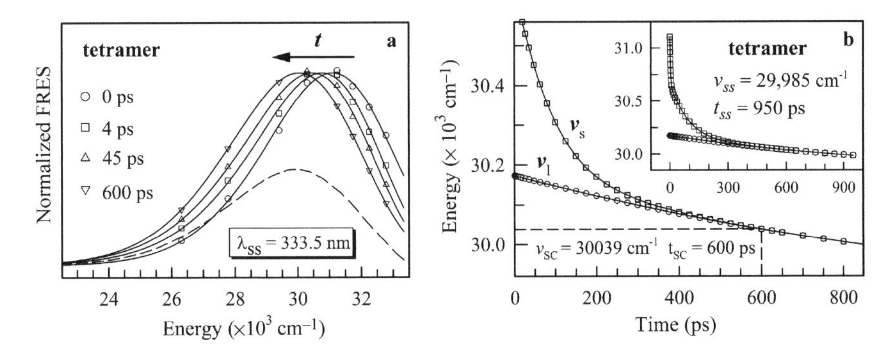


**Figure 11.** (a) Normalized overall FRES of melittin tetramer at several representative times. The dashed curve is the steady-state emission spectrum with the peak at 333.5 nm. (b) fs-resolved emission maxima of the overall FRES ($\nu_s$) and lifetime-associated FRES ($\nu_l$) of melittin tetramer. The inset shows the entire evolution of $\nu_s$ and $\nu_l$ to reach the steady-state emission ($\nu_{ss}$).

wavelength-resolved fluorescence temporal behaviors in the tetramer conformation. We observed significantly slow solvation at the blue side, and the two solvation components decay in 2–10 and 30–100 ps. At the red side ($\geq$335 nm), we also clearly observed two initial rise components in ~0.5 and 5 ps. The constructed FRES is shown in Fig. 11a, and the spectrum shifts from solvation until about 600 ps, which is significantly different from free tryptophan in water in Fig. 3b, as also shown for $\nu_s(t)$ and $\nu_l(t)$ in Fig.11b and Fig. 3c. The final solvation response curves for three environments are given in Fig. 12a along with their anisotropy dynamics in Fig. 12b.

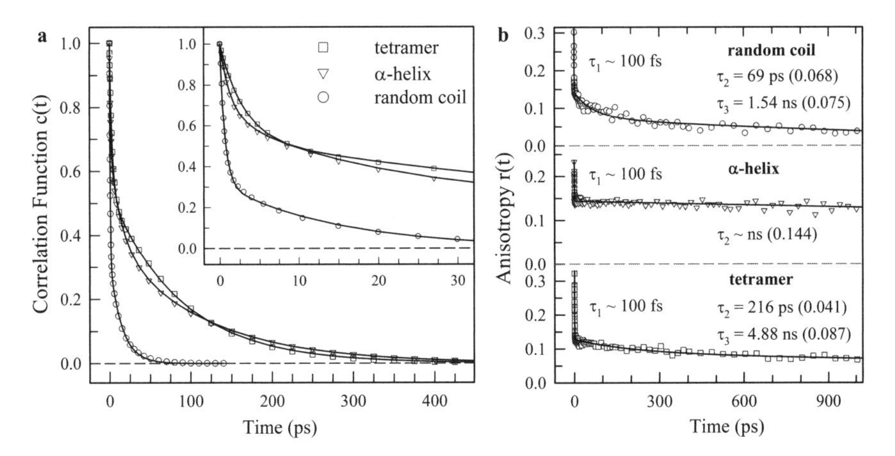


**Figure 12.** (a) Hydration correlation functions $c(t)$ of melittin probed by tryptophan in three different conformations. (b) fs-resolved fluorescence anisotropy dynamics of tryptophan in three conformations of melittin. The long component represents protein tumbling motion.

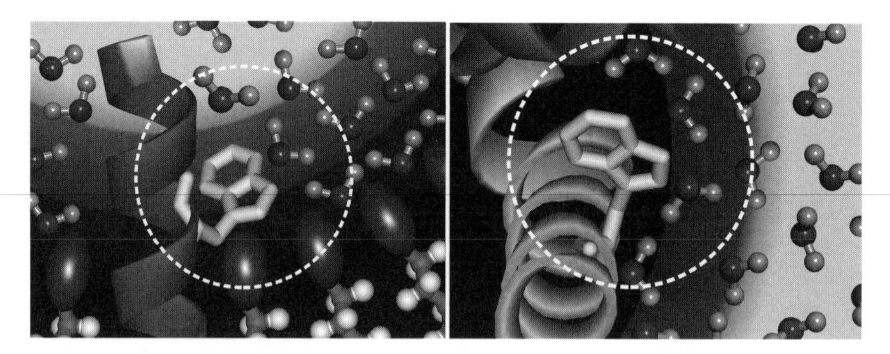

**Figure 13.** Schematic representation of the local ordered water network in the α-helical melittin (left) and melittin tetramer (right). The probe sensitivity is indicated by the dashed circle in a diameter of ~7–10 Å. Different layers of water structures are represented with circles from dark to light.

With femtosecond resolution, we observed the solvation dynamics that occur in 0.62 ps (68%) and 14.7 ps (32%) in a random-coiled primary structure. The former represents bulk-like water motion, and the latter reflects surface-type hydration dynamics of some proteins [45, 46]. Clearly, even for a random coil, a certain ordered, rigid water network has been formed. At a membrane–water interface, as schematically shown in Fig. 13, melittin folds into a secondary α-helical structure, and the tryptophan senses three different-type water motions in the water nanochannel (also see Section III.D) on the time scales of 1.25 ps (35%), 13.9 ps (27%), and 114 ps (38%). The interfacial water motion between lipid and channel water was found to take as long as 114 ps, which indicates a well-ordered, rigid water structure along the membrane surface. In a high-salt aqueous solution, the dielectric screening and ionic solvation promote the hydrophobic core collapse in melittin aggregation and facilitate the tetramer formation. This self-assembled tertiary structure is also stabilized by the strong hydrophilic interactions of charged C-terminal residues and associated ions with water molecules in the two assembled regions. The hydration dynamics (Fig. 12) was observed to occur in 3 ps (47%) and 87 ps (53%), which is significantly slower than water relaxation around the random-coil surface but similar to water motion at membrane interfaces. Thus, the observed time scale of ~100 ps of water motion probably implies appropriate water mobility for mediating formation of high-order structures of melittin in an α-helix and a self-assembled tetramer. The observed anisotropy dynamics (Fig. 12b), which reflect local rigidity around the probe, have no *direct* correlation with the time scales of hydrating water motions. These results elucidate the critical role of hydration dynamics in peptide conformational transitions and protein structural stability and integrity (Fig. 9).

## C. Drug Transporter Human Serum Albumin

Biomolecular recognition is mediated by water motions, and the dynamics of associated water directly determine local structural fluctuation of interacting partners [4, 9, 91]. The time scales of these interactions reflect their flexibility and adaptability. For water at protein surfaces, the studies of melittin and other proteins [45, 46] show water motions on tens of picoseconds. For trapped water in protein crevices or cavities, the dynamics becomes much slower and could extend to nanoseconds [40, 71, 92]. These rigid water molecules are often hydrogen bonded to interior residues and become part of the structural integrity of many enzymes [92]. Here, we study local water motions in various environments, from a buried crevice to an exposed surface using site-selected tryptophan but with different protein conformations, to understand the correlation between hydration dynamics and conformational transitions and then relate them to biological function.

Human serum albumin (HSA) is an important transporter of fatty acids, metabolites, drugs, and organic compounds in the circulatory system [93, 94]. It is a single polypeptide chain consisting of 585 amino acids. Under physiological conditions (pH ~7), HSA adopts a heart-shaped three-dimensional (3D) structure with three homologous domains I–III (Fig. 14); each domain contains two subdomains A and B, which consist of four and six α-helices, respectively [95, 96]. The X-ray structure shows that two halves of the albumin molecule

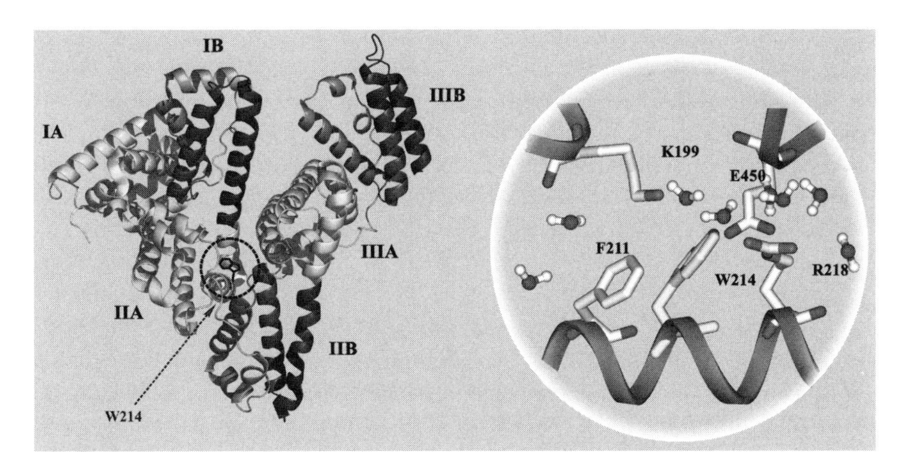


**Figure 14.** Left: X-ray crystallographic structure of human serum albumin at the neutral pH condition (PDB ID: 1N5U). The protein is constituted with three domains, I, II, and III, and each domain has two subdomains, A (light gray) and B (dark gray). Note a deep crevice between the two halves of the structure with W214 (in black sticks) located at the bottom. Right: Local configuration around W214 with eight trapped water molecules within 7 Å and four potential tryptophan fluorescence quenching residues in close proximity of less than 5 Å.

form a 10-Å wide, 12-Å deep crevice, which is filled with water molecules [95, 96]. The only single tryptophan residue W214 in the protein is located in the binding site IIA at the bottom of the crevice (Fig. 14).

With a change in pH, HSA undergoes reversible conformational isomerization, and the pH-dependent isomers were first demonstrated by Luetscher in 1939 [97] and later systematically classified by Forster in 1960 [98]. At a neutral pH, the conformation of HSA is in its common physiological state, which is referred to as a norm form (**N**) with its emission maximum at 338 nm. This form is similar to the surface-tryptophan emission [45, 46, 70], which indicates a highly polar environment with a large amount of trapped water around W214 in the crevice. An abrupt transition occurs at a pH value of less than 4.3, which changes the **N** form to the so-called fast migrating form (**F**) with the emission peak shifting to the blue at 332.6 nm, suggesting a less polar microenvironment. When the pH is less than 2.7, another transition takes place from the **F** form to the extended form (**E**), which leads to another blue shift of emission maximum to 330 nm. On the other side of the scale, when the pH value becomes above 8, the **N** conformation changes to the basic form (**B**) with an emission spectrum peaking at 336 nm, which indicates a slight change of local polarity. At a pH above 10, the structure transforms to another aged form (**A**) without further shift of its emission spectrum. These transitions are summarized below [93].

$$\begin{array}{l} \text{Conformation}\ \mathrm{E} \underset{2.7}{\longleftrightarrow} \mathrm{F} \underset{4.3}{\longleftrightarrow} \mathrm{N} \underset{8}{\longleftrightarrow} \mathrm{B} \underset{10}{\longleftrightarrow} \mathrm{A} \\ \text{pH transition} \end{array}$$

The protein has been studied extensively in every aspect because of its physiological importance and its potential as a drug delivery vehicle [93, 94], and because different acidic and basic isomers may have certain biological function [90, 100]. Most studies have focused on binding interactions with various ligands and have revealed that HSA is an assembly of squirmy and resilient components that frequently change in conformation through opening and closing of major crevices. With this "breathing" motion, HSA assimilates and releases a variety of substances during transportation in the circulation. Many complex structures have also been resolved at high X-ray resolution [95, 101, 102]. The flexibility to adopt different conformations with independent segmental movements has also been shown through certain independent and sequential folding/unfolding of individual subdomains [103, 104]. The tryptophan W214 and extrinsic labeled dye molecules have been used as optical probes to study the conformational dynamics through changes in fluorescence lifetime, resonance energy transfer, and time-resolved anisotropy [105–109]. The significant change in the lifetime of W214 for a series involving mutation in the binding site IIA clearly showed the local flexibility [110].

We examined the ligand-binding interactions of HSA in earlier studies [111], elucidating the rigid hydrophobic recognition. Here, we report a systematic investigation of the solvation dynamics of tryptophan (W214) in different conformations of HSA with the aim of understanding the dynamics and structural flexibility under physiological conditions (**N** form) and the dynamical changes for conformational transitions that occur at different pH values. Because it is known that water is present (eight water molecules shown in Fig. 14) around tryptophan in the crevice of the **N** conformation, the dynamics should reflect the influence of their motions. Figure 15 shows the fs-resolved fluorescence transients of W214 for several typical wavelengths in **N** conformation. The overall decay dynamics is significantly slower than that of aqueous tryptophan in a similar buffer solution [49]. Clearly, the ultrafast decay components ($<1$ ps) observed in tryptophan solution were not observed at the blue side for all isomers. The two exponential decays of solvation have time constants of 4.4–9.2 ps and 104–125 ps for all the blue-side transients. At the red side, we also observed a minor decay component of 51 ps with $\sim$10% of the total amplitude, which results from the excited-state quenching of W214 by the neighboring residue(s) [110]. Similar slow dynamics

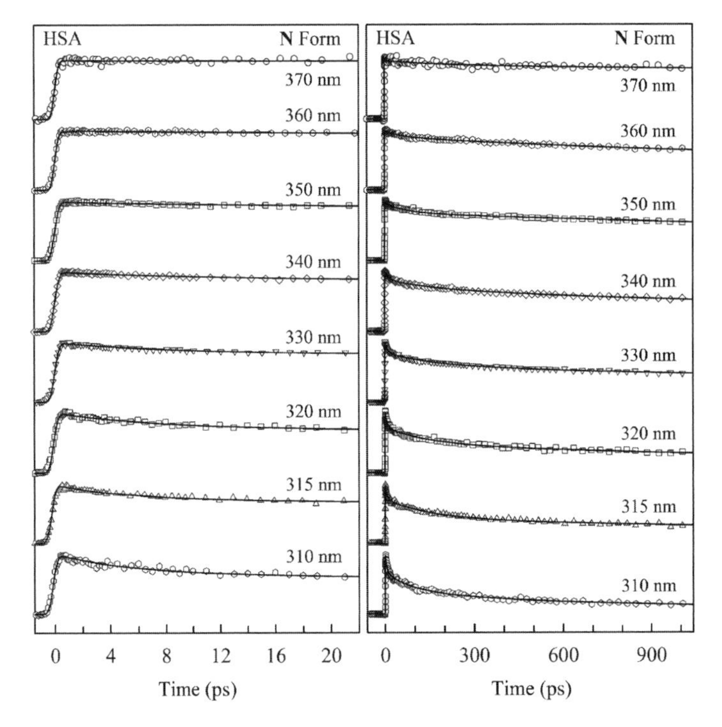


**Figure 15.** Normalized, fs-resolved fluorescence transients of the **N** isomer of HSA in the short (left) and long (right) time ranges, with a series of gated fluorescence emissions.

of the fluorescence transients were observed for **F** form in 4.1–10.7 ps and 125–208 ps and **E** form in 6.7–16.2 ps and 72–120 ps. However, significant fast solvation dynamics were observed in 1.1–3.1 ps and 32–45 ps for **B** form and 1.1–3.5 ps and 15–28 ps for **A** form.

The constructed correlation functions of these isomers are given in Fig. 16a along with their anisotropy dynamics in Fig. 16b. All correlation functions show a robust double-exponential decay: **N** form, 5.0 ps (39%) and 133 ps (61%); **F** form, 4.3 ps (30%) and 186 ps (70%); **E** form, 6.7 ps (39%) and 108 ps (61%); **B** form, 1.6 ps (30%) and 46 ps (70%); and **A** form, 2.3 ps (76%) and 27 ps (24%). Significantly, we observed the most flexible local conformation in the physiological **N** form among all isomers (Fig. 16b) with a wobbling angle of 24°. These observations of solvation dynamics and local flexibility in a series of reversible conformations of human serum albumin are significant. For all conformational isomers, the initial solvation dynamics occurs in several picoseconds, which reflects local reorientation motions of surrounding trapped water molecules and possibly the associated polar/charged residues. The observed second and longer time solvation dynamics occurs in the range of tens to hundreds of picoseconds, depending on conformer, and reflects the site-hydrated water network rearrangements in the crevice. The Stokes shift is an integration of solvation processes, whereas solvation

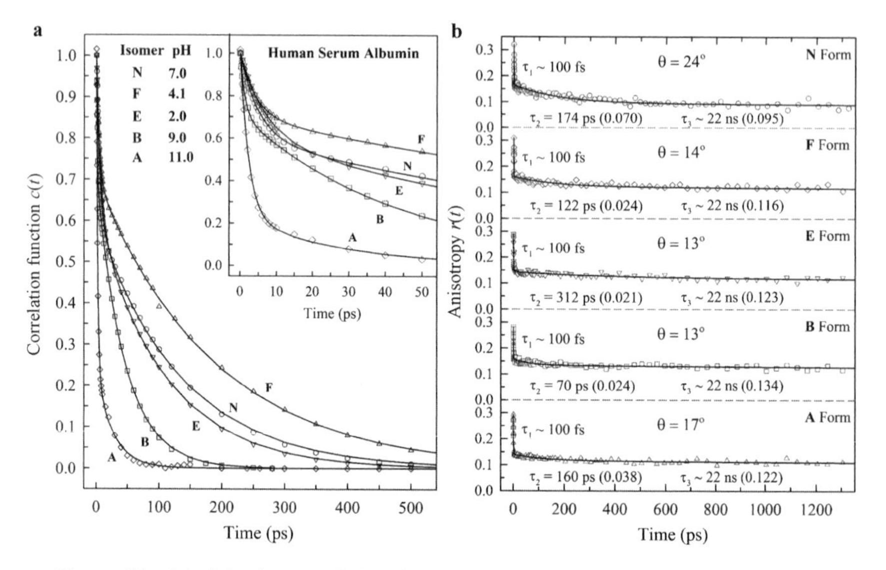


**Figure 16.** (a) Solvation correlation functions probed by tryptophan (W214) in the five conformational isomers of HSA. The inset shows the correlation functions in the short-time range. (b) fs-resolved fluorescence anisotropy dynamics of W214 of the five conformational isomers. The derived wobbling motion is also given in terms of the semiangle ($\theta$).

dynamics measure the rate of local hydrogen-bond rearrangements. This long time scale depends on the local water structure in the crevice, which is known to be eight in number (Fig. 14), and local chemical identity. Depending on local rigidity, polar/charged-protein residues may contribute to the observed total solvation dynamics, but given the conformational and orientational correlations reported here, hydration by the smaller water molecules is major, as water molecules are already present in the crevice and will contribute solely or in association with residues to solvation. The result reported here for the shortening of this longer time solvation, when HSA opens up its structure (conformer of basic pH), is consistent with the trapped water being more mobile.

Under the normal physiological condition of neutral pH, the single W214 tryptophan in the binding site IIA is buried inside a 10-Å deep crevice. Despite this closure, we observed fast solvation dynamics of 133 ps, which revealed a binding pocket capable of solvation on the fast time scale. Moreover, this globular native structure of the normal isomer shows the largest flexibility among all conformational isomers as revealed from measurement of the local wobbling motions of W214 tryptophan. The large plasticity of HSA is essential for the albumin molecule to accommodate a variety of ligands and to perform the transport function in the circulation. The observed time scale of $\sim$100 ps is ideal for the trapped water molecules in the crevice at the binding site to maintain the local structural integrity as well as to maintain dynamic flexibility and lubrication in recognition and conformational transitions (Fig. 17). The "lock-and-key or induced-fit" concept of molecular recognition is therefore incomplete without knowledge of dynamical solvation and plasticity of the protein.

### D. Lipid Interface and Nanochannel

Water confined in self-organized molecular assemblies is of great interest to biology and materials [4, 9, 112–118]. Examples of water molecules in such restricted environments are abundant not only at surfaces [45, 46, 70, 72–74] and in the interior of proteins [71, 119–122] as we reviewed above, but also in the grooves of DNA [123–126], on micelles or in water pools of reverse micelles [127–135], and in porous materials [114, 136–138]. Elucidation of these confined water structures and dynamics is central to understanding the biological functions and material properties of their associated macromolecules in general. A variety of spectroscopic techniques [45, 70, 127, 139–141] has been used to characterize these water structures. Very recently, a unique water structure with anomalously soft dynamics inside single-walled carbon nanotubes has been reported [142]. Here, we use a model system of water nanochannels formed in the lipidic cubic phase (Pn3m) and systematically examine water dynamics across the channel.

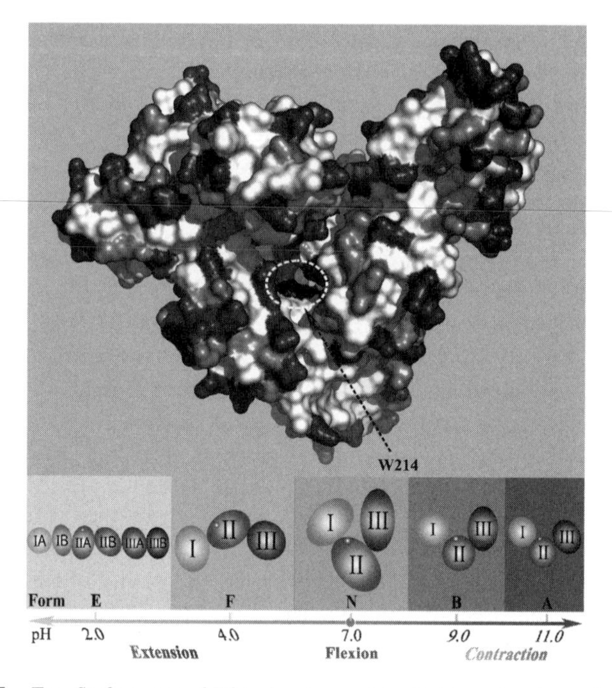


**Figure 17.** Top: Surface map of HSA showing positive (light gray) and negative (dark gray) charge distributions. The crevice is shown in the center with tryptophan (black patch in white circle) sitting at its bottom. Bottom: Schematic representation of conformational transitions, from the contracted configuration in basic pH, to the flexible globule structure at neutral pH, and to the extended form in acidic pH. The location of W214 is indicated by the white dot in domain IIA.

The lipidic cubic phase has recently been demonstrated as a new system in which to crystallize membrane proteins [143, 144], and several examples [143, 145, 146] have been reported. The molecular mechanism for such crystallization is not yet clear, but the interfacial water and transport are believed to play an important role in nucleation and crystal growth [146, 147]. Using a related model system of reverse micelles, drastic differences in water behavior were observed both experimentally [112, 127, 128, 133–135] and theoretically [117, 148, 149]. In contrast to the ultrafast motions of bulk water that occurs in less than several picoseconds, significantly slower water dynamics were observed in hundreds of picoseconds, which indicates a well-ordered water structure in these confinements.

An amphiphilic fluorescence probe with a variable alkyl chain length to tether to lipid bilayers was used. The chromophore is an indole moiety, and the probes are a series of Trp-alkyl ester molecules. Presumably, the hydrophobic alkyl chains systematically anchor the chromophore to different depths of the

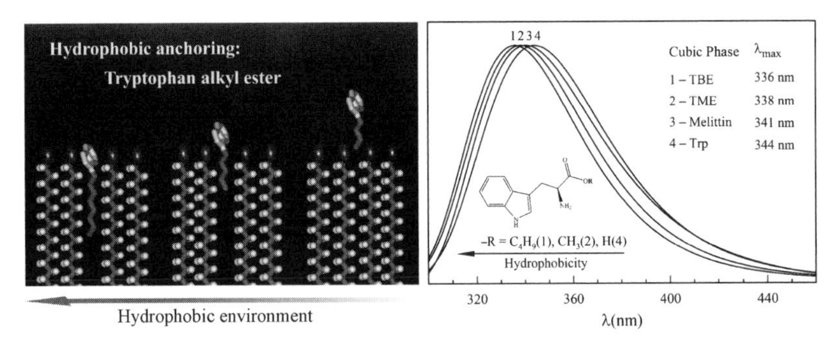


**Figure 18.** Left: Schematic representation of the concept for probing different layers of water near lipid–water interfaces through anchoring hydrocarbon tails of a series of Trp-alkyl ester probes into lipids. Right: Normalized steady-state fluorescence emissions from four Trp probes. Note the correlation between emission maxima and their hydrophobicity.

lipid bilayer, which enables it to sample various aqueous environments and probe different layers of water dynamics, which is conceptually shown in Fig. 18. With changes in alkyl length from tryptophan, to tryptophan methyl ester (TME) and to tryptophan butyl ester (TBE), we systematically anchored the hydrocarbon tails of the probes into lipids with increase in hydrophobicity (Fig. 18) and examined the water structures across the nanochannel. The solvation dynamics probed by melittin interacting with lipids in the cubic phase, which are shown in Fig. 12a, are also included in the series.

Figure 19 shows the typical fluorescence transients of TBE from more than 10 gated emission wavelengths from the blue to the red side. At the blue side of the emission maximum, all transients obtained from four Trp-probes in the cubic phase aqueous channels drastically slow down compared with that of tryptophan in bulk water. The transients show significant solvation dynamics that cover three orders of magnitude on time scales from sub-picosecond to a hundred picoseconds. These solvation dynamics can be represented by three distinct decay components: The first component occurs in about one picosecond, the second decays in tens of picoseconds, and the third takes a hundred picoseconds. The constructed hydration correlation functions are shown in Fig. 20a with anisotropy dynamics in Fig. 20b. Surprisingly, three similar time scales (0.56–1.431 ps, 9.2–15 ps, and 108–140 ps) are obtained for all four Trp-probes, but their relative amplitudes systematically change with the probe positions in the channel. Thus, for the four Trp-probes studied here, we observed a correlation between their local hydrophobicity and the relative contributions of the first and third components; from Trp, melittin, TME to TBE, the first components have contributions of 40%, 35%, 26%, and 17%, and the third components vary from 32%, to 38%, 43%, and 53%, respectively. The

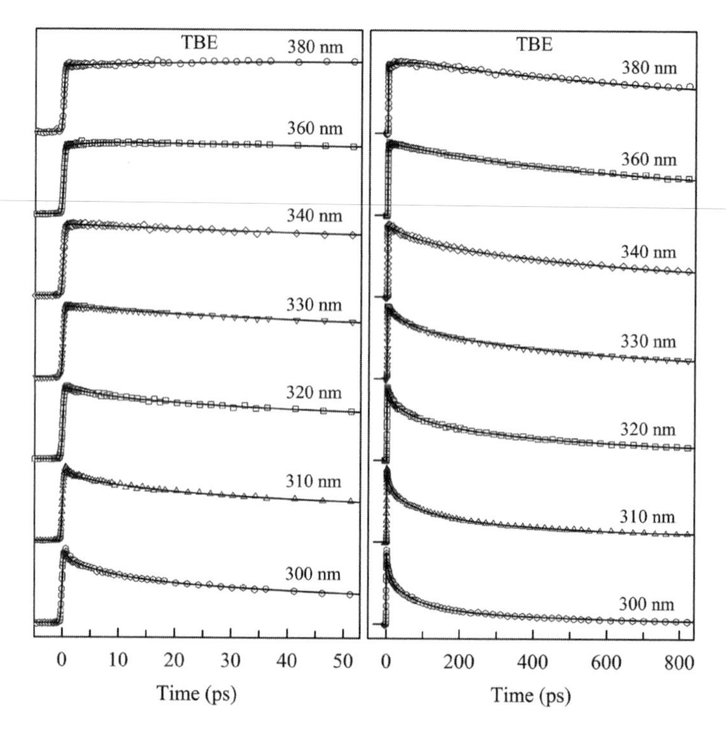


**Figure 19.** Normalized fs-resolved fluorescence transients of TBE in the aqueous channels of the cubic phase in the short (left) and long (right) time ranges with a series of gated fluorescence emissions.

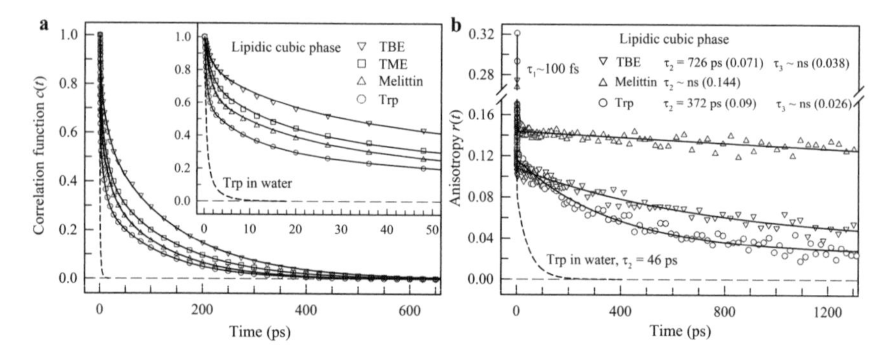


**Figure 20.** (a) Hydration correlation functions $c(t)$ obtained for four Trp probes (solid lines). The inset shows the correlation functions in the short time range. The hydration correlation function of tryptophan in bulk water is also shown (dashed line) for comparison. (b) fs-resolved fluorescence anisotropy dynamics of Trp probes in the aqueous channels of the cubic phase. For clarity, the anisotropy of TME is not shown. The relaxation of Trp in bulk water (46 ps) is also shown (dashed line) for comparison.

second components have nearly constant amplitudes of about 30%. The observed solvation dynamics of channel water are striking. The three distinct time scales and the systematically varying amplitudes, which are well correlated with their hydrophobicity, strongly suggest discrete water structures in the nanochannel. These three types of water structures represent bulk-like, quasi-bound, and well-ordered water distributions across the aqueous channel (Fig. 21).

Extensive studies in reverse micelles revealed a similar water distribution [127–130], which is consistent with the distinct water model proposed by Finer [150]. For example, when the molar ratio ($w_0$) of water to the surfactant is 6.8 in lecithin reverse micelles with a corresponding diameter of ~37 Å, three solvation time scales of 0.57 (13%), 14 (25%), and 320 ps (62%) were observed using coumarin 343 as the molecular probe. At $w_0 = 4.8$ with a ~30-Å water core diameter, only a single solvation dynamic was observed at 217 ps, which indicates that all water molecules are well ordered inside the aqueous pool. The lecithin in these reverse micelles have charged headgroups, which have much stronger interactions with water than the neutral headgroups of monoolein in the

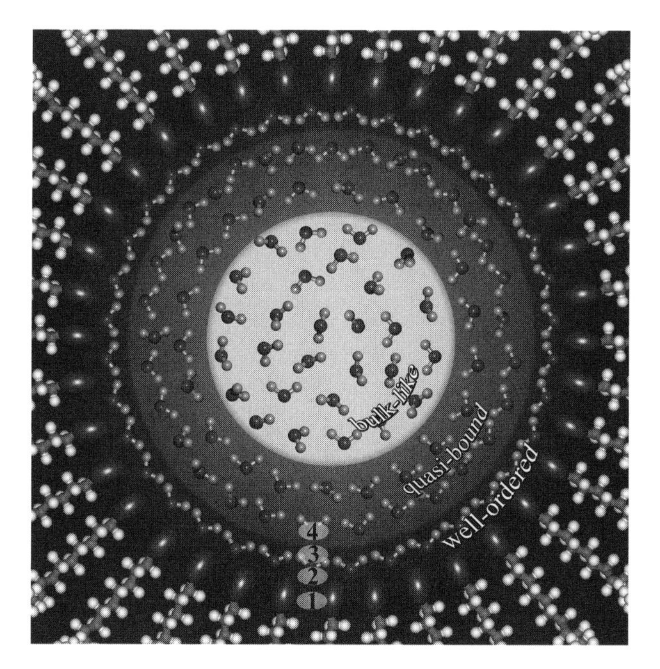


**Figure 21.** Schematic representation of water structure in aqueous nanochannels ~50 Å in diameter. By anchoring four Trp probes into lipid bilayers, these molecular rulers measure water motions in different local regions in the channel as shown in positions 1–4 of TBE, TME, melittin, and Trp.

cubic phase channels. The longest solvation component (∼200–300 ps) represents water relaxation near the charged headgroups at the lecithin interface and is expected to be slower than that of monoolein interfacial channel water (∼100 ps). In the latter case, the channel diameter is about 50 Å and is equivalent to $w_0 \approx 11.4$ in reverse micelles. Thus, there is considerable free bulk water at the core of the nanochannel.

The response range of the local environment to the excited Trp-probe is mainly within 10 Å because the dipole–dipole interaction at 10 Å to that at ∼3.5 Å of the first solvent shell drops to 4.3%. This interaction distance is also confirmed by recent calculations [151]. Thus, the hydration dynamics we obtained from each Trp-probe reflects water motion in the approximately three neighboring solvent shells. About seven layers of water molecules exist in the 50-Å channel, and we observed three discrete dynamic structures. We estimated about four layers of bulk-like free water near the channel center, about two layers of quasi-bound water networks in the middle, and one layer of well-ordered rigid water at the lipid interface. Because of lipid fluctuation, water can penetrate into the lipid headgroups, and one more trapped water layer is probably buried in the headgroups. As a result, about two bound-water layers exist around the lipid interface. The obtained distribution of distinct water structures is also consistent with ∼15 Å of hydration layers observed by X-ray diffraction studies from White and colleagues [152, 153]. These discrete water structures in the nanochannel are schematically shown in Figure 21, and these water molecules are all in dynamical equilibrium.

The TBE and the tryptophan octyl ester (TOE) in the cubic phase give the same fluorescence emission maximum at 336 nm, which indicates that both TBE and TOE insert into the lipid headgroups (Fig. 18) but are not buried in the lipid hydrocarbon chains because tryptophan in a truly hydrophobic environment has an emission maximum at less than 330 nm. Also, TBE is not located at the lipid surface because the surface tryptophan has a maximum close to ∼340 nm. We also measured the anisotropic dynamics, and the result (Fig. 20b) is consistent with highly restricted wobbling motions by lipid headgroups with a long decay time constant of 726 ps. TBE in water takes less than 100 ps to complete rotational relaxation. More importantly, the anisotropy finally stays at a constant within 1.5 ns after partial orientational relaxation, which shows that TBE experiences a hindered local motion by the lipid headgroups. Recent experiments and MD simulations [154] in micelles also supported the insertion position of TOE into the headgroups. Thus, the observed 147-ps solvation dynamics mainly result from the trapped and ordered water motion around the lipid interface. The contribution from the headgroups would be minor [145, 149]. The 15-ps hydration relaxation reflects the quasi-bound water motion near the lipid interface, and this time scale is very similar to that of the surface water motion of proteins. The ultrafast component in 1.43 ps can be from bulk-like water relaxation as well as from the initial libration of ordered-water molecules. According to the location of TBE in the channel (position 1 in Fig. 21) and the sensitivity of the Trp-probe, the observed ultrafast dynamics favors the latter contribution of libration. The overall orientation of well-ordered water at the lipid interface, although in a fluctuation on the time scale of ~150 ps, aligns along the lipid headgroup polarization, which results in its dipole direction pointing to the channel center (Fig. 21), as is confirmed by recent MD simulations [155].

With the probe position moving toward the center of the aqueous channel, we detected more ultrafast and less slow solvation components. Note the negligible change of the quasi-bound water contributions, which indicates the complete detection of the two layers of quasi-bound water by all four Trp-probes. For TME, the fluorescence emission peak shifts to 338 nm, and its location moves to the lipid interface (Fig. 18). We did observe a smaller fraction of slow solvation dynamics decreasing from 53% in TBE to 43% in TME and an increase of the ultrafast component from 17% to 26%. The corresponding anisotropy dynamics drops from 726 to 440 ps with a less hindered local motion at the lipid interface.

The emission of Trp19 in melittin shifts to the red side peaking at 341 nm (Fig. 18), and the probe location slightly moves away from the lipid interface toward the channel center. Consistently, we observed a larger fraction of the ultrafast solvation component (35%) and a smaller contribution of slow ordered-water motion (38%). Melittin consists of 26 amino acid residues (Fig. 9), and the first 20 residues are predominantly hydrophobic, whereas the other 6 near the carboxyl terminus are hydrophilic under physiological conditions. This amphipathic property makes melittin easily bound to membranes, and extensive studies from both experiments [156–161] and MD simulations [162–166] have shown the formation of an $\alpha$-helix at the lipid interface. Self-assembly of $\alpha$-helical melittin monomers is believed to be important in its lytic activity of membranes [167–169]. Our observed hydration dynamics are consistent with previous studies, which support the view that melittin forms an $\alpha$-helix and inserts into the lipid bilayers and leaves the hydrophilic C-terminus protruding into the water channel. The orientational relaxation shows a completely restricted motion of Trp19, and the anisotropy is constant in 1.5 ns (Fig. 20b), which is consistent with Trp19 located close to the interface around the headgroups and rigid well-ordered water molecules.

Tryptophan emission in the channel shifts another 3 nm to the red and peaks at 344 nm (Fig. 18), and it has a maximum at ~350 nm in bulk water. Thus, the probe location moves toward the channel center, but it is situated in the quasi-bound water layer (Fig. 21), which is consistent with the obtained hydration results. Again, we observed an increase in the ultrafast component (40%) and the decrease of the slow contribution (32%). The observed 108-ps slow dynamics are still from the ordered-water motion at the interface and the 9.2-ps component from the quasi-bound water relaxation. The 0.56-ps dynamics represents ultrafast bulk-like water motion and local libration, which indicates that a significant fraction of bulk-like water ($\leq$40%) was probed. The fs-resolved anisotropy showed a local restricted motion with a relaxation time of 372 ps (Fig. 20b), one order of magnitude slower than in bulk water (46 ps), which reveals local rigid water networks and is consistent with aforementioned hydration results.

The observed discrete water dynamics and structures in the nanochannel are significant and probably impact on the nanoscopic interactions of biomolecules with water and lipids. The ultrafast bulk-like water motion in less than 1 ps freely transports biomolecules near the channel center, which is consistent with the related transport studies [147]. The well-ordered water at the lipid interface facilitates anchoring of biomolecules in the lipid bilayer with dynamical rigidity on the time scale of hundreds of picoseconds. The quasi-bound water near the lipid interface fluctuates in tens of picoseconds and keeps biomolecular conformational flexibility for biological function. Water is so unique in nano-confinements and can adopt various structures and dynamical motions for performing different biological functions and having exceptional properties of nanomaterials.

## E. Supramolecule Crown Ether

Crown ethers are heteromacrocycles in which the framework is typically composed of repeating ethyleneoxy [-$(CH_2CH_2O)_n$-] units. Nitrogen and sulfur commonly replace oxygen in this framework, which leads to a great variety of compounds that have been used in molecular recognition studies and supramolecular chemistry [170–175]. Crown ethers form more or less stable complexes in the solution and vapor phases with a variety of organic and metallic cations. In these host–guest recognition processes, solvent plays a critical role in local structure optimization and complex stabilization. Thus, complex stability is known to vary, sometimes drastically, according to the solvent in which the reaction occurs [176]. Understanding of supramolecule-solvent interactions and structures at the local molecular level is important for potential applications in drug delivery, and chemical sensor development, and to supramolecular chemistry in general.

Crown ethers selectively complex various alkali metal cations and can be thus used as model systems to study interactions between a macrocycle-bound cation and the $\pi$-system of a sidearm arene. Alkali-metal cation-$\pi$ interactions have recently received considerable attention because of the biological importance [88, 89, 175]. These studies have focused on $Na^+$ and $K^+$ interacting with benzene, phenol, and indole, which are the side chain arenes of phenylalanine, tyrosine, and tryptophan, respectively. Recent work [177–180] has demonstrated the formation of stable complexes between, for example, $K^+$

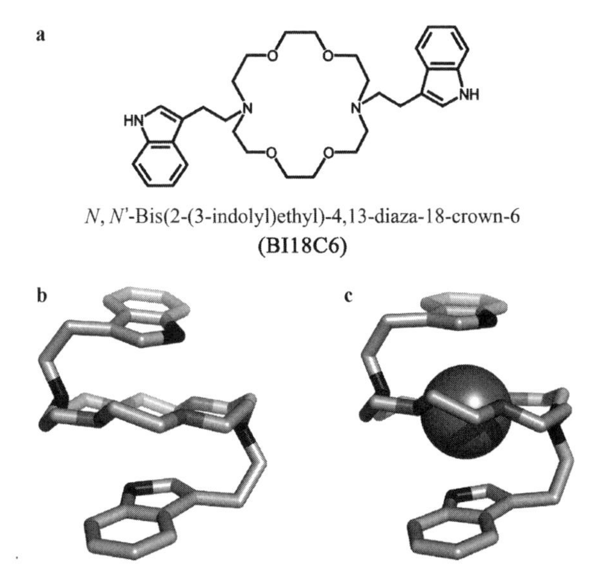


**Figure 22.** (a) Molecular structure of **BI18C6**. (b) **BI18C6** structure in solvent in tubular representation. (c) Solid-state X-ray structure of **BI18C6** complexed with a potassium cation $K^+$ in tubular representation. The $K^+$ is shown as a big ball.

and a crown ether in which two side-armed indole rings are in contact with the macroring-bound cation (see Fig. 22c). The complex structure was obtained by the solid-state characterization using X-ray diffraction [177, 178] and later by NMR studies in solution [181, 182]. However, the structure without cation recognition was believed to be stretched, not folded.

We studied two arene side-armed lariat ether receptors in the presence and absence of $K^+$, in two common solvents: acetonitrile ($CH_3CN$) and methanol ($CH_3OH$). The two crown ethers used were *N*, *N'*-Bis(2-(3-indolyl)ethyl)-4,13-diaza-18-crown-6 (**BI18C6** in Fig. 22a), which has two indolylethyl sidearms; and *N*-(2-(3-indolyl)ethyl)aza-18-crown-6, which has a single indolylethyl side chain. The former was chosen as the host supramolecule for recognition study, and the latter was studied for comparison. The indole moiety in these molecules was used as an optical probe to examine the solvent relaxation dynamics around the supramolecule with and without encapsulation of $K^+$. Figure 23 shows steady-state fluorescence spectra of tryptophan and **BI18C6** in methanol and acetonitrile with 290 nm excitation. The emission peak of tryptophan in water is at 349 nm [49], and the peak shifts to the blue side at 336.5 nm in methanol and 332.3 nm in acetonitrile. Surprisingly, the emission peak in acetonitrile from the solute tryptophan to **BI18C6** shifts to the red side at 350.5 nm by 1563 $cm^{-1}$. This observation is striking and quite unusual. The macroring of the crown ether

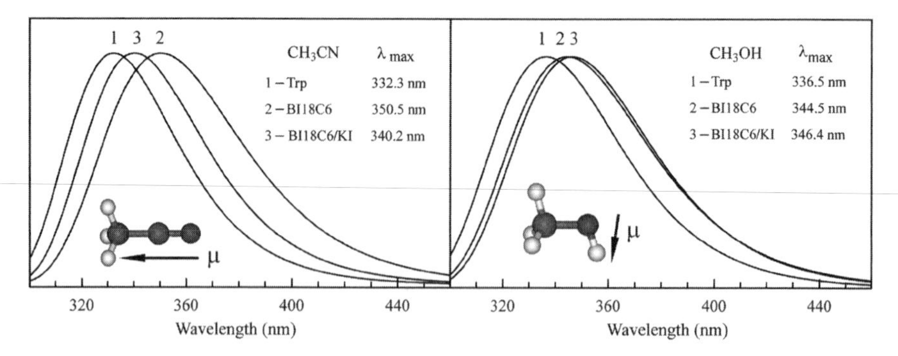


**Figure 23.** Steady-state fluorescence spectra of tryptophan and supramolecule **BI18C6** in $CH_3CN$ and $CH_3OH$ at excitation of 290 nm. The ball-stick structures of two solvent molecules with their dipole moments are also shown. Note the different shifts in two solvents with and without addition of KI.

does not significantly change the electronic structures of the indole moiety. The observed large shift must be from the solvent effect; the solvent molecules organize around the macroring of crown ethers and a well-ordered structure is formed, which results in enhancement of the local polarization. With the addition of KI, the emission peak shifts back to the blue side at 340.2 nm. This observation indicates that the supramolecule-solvent structure is probably different with and without KI. With the encapsulation of $K^+$, the recognition leads to local solvent rearrangements and thus results in a decrease of the local polarization and a blue shift of the emission. The addition of KCl gave a similar emission maximum at 340.6 nm. In methanol, the emission peak shifts to 344.5 nm by 690 $cm^{-1}$ when changing tryptophan to **BI18C6**, which again reflects an ordered local solvent structure around the macroring, but with less enhanced local polarization than in acetonitrile. With the addition of KI, the peak slightly shifts to the more red side at 346.4 nm because of the encapsulation of $K^+$. All observed spectral shifts from tryptophan to **BI18C6** are related to ordered supramolecule-solvent structures and solvent dipole moments where acetonitrile is 3.92 D whereas methanol is 1.70 D. The ordered local solvent structures around the crown ether are studied directly by the microsolvation dynamics below.

Figure 24 shows three typical fs-resolved fluorescence transients of **BI18C6** without and with the addition of KI in acetonitrile from more than 10 different gated emissions that cover from 305 nm to 440 nm. All the transients show the typical solvation relaxation behavior with initial decay in the blue-side emission and rise at the red side. However, with and without encapsulation of $K^+$, the solvation dynamics are drastically different. For example, at 305 nm, the solvation decays without $K^+$ occur in 0.39 ps, 3.6 ps, and 105 ps but with $K^+$ the transient

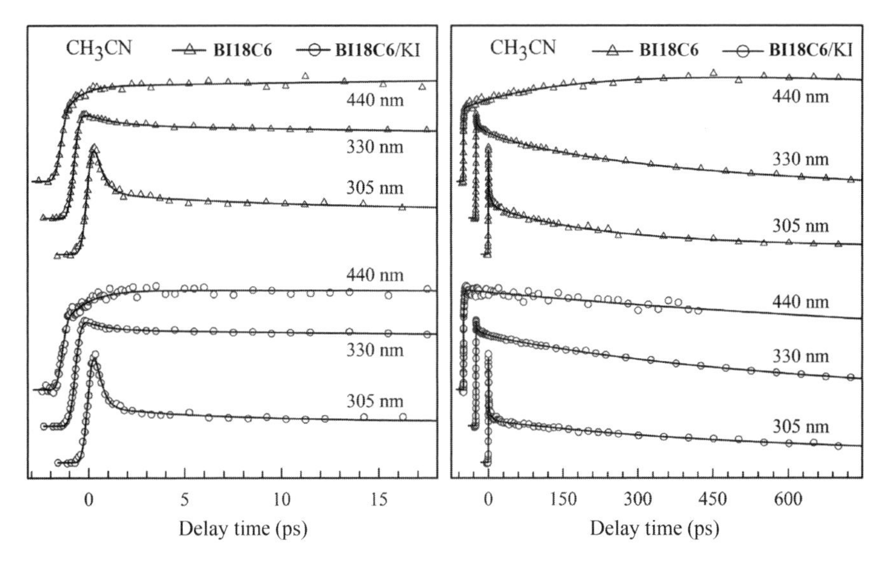


**Figure 24.** Normalized, fs-resolved fluorescence transients of the supramolecule **BI18C6** in $CH_3CN$ without and with the addition of KI in the short (left) and long (right) time ranges from more than 10 gated fluorescence emissions. Note the drastic difference of transients at the long delay time without and with encapsulation of the cation $K^+$.

decays with 0.32 ps and 4.6 ps. The long solvation component of 105 ps disappeared by adding KI into the solution. At the far red-side emission of 440 nm, without $K^+$ we observed two initial rise components with time constants of 0.84 ps and 261 ps. With $K^+$, the transient initially rises in 1.2 ps. Clearly, in the absence of KI, one slow solvation component in hundreds of picoseconds appears in all transients, which indicates a well-ordered solvent structure around the supramolecule. With encapsulation of $K^+$, this long component is absent in all measured transients, which is indicative of rearrangements of solvent molecules during recognition with a final less ordered structure.

We also systematically measured the fs-resolved fluorescence transients with the addition of KCl, and the results are nearly the same as those obtained with KI. The anion $Cl^-$ and $I^-$ make no difference to the solvation dynamics observed here. As a comparison, we also studied another crown ether supramolecule with only one indole sidearm. With and without KI, the results are similar to those obtained for **BI18C6** with two indole sidearms. This observation is consistent with the folded symmetric structure (Fig. 22b and c), and either of two indole rings gives similar solvation behaviors. However, in solvent $CH_3OH$, the solvation dynamics, decay in 0.95–16.3 ps and rise in 0.55–12 ps, are similar in the presence and absence of $K^+$ with slightly faster processes for the former. Both results have the similar trend as tryptophan in $CH_3OH$.

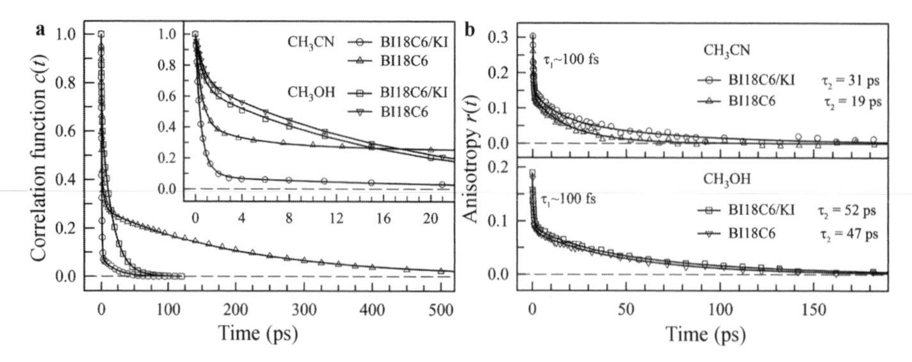


**Figure 25.** (a) Solvation correlation functions c(t) of $CH_3CN$ and $CH_3OH$ probed by the indole moiety of the supramolecule **BI18C6** with and without encapsulation of the cation $K^+$. The inset shows the correlation functions in the short time range. For clarity, the correlation function of $CH_3OH$ probed by **BI18C6** is not shown in the long time range. (b) fs-resolved fluorescence anisotropy dynamics of **BI18C6** in two solvents and note the similar time scales with and without encapsulation of the cation $K^+$.

The final solvent correlation functions are shown in Fig. 25a along with anisotropy dynamics in Fig. 25b. One drastically distinct correlation function is from **BI18C6** in acetonitrile without cation complexation in 0.48 ps (55%), 3.9 ps (15%), and 206 ps (30%). With the addition of KI, the solvation dynamics become faster with 0.4 ps (76%), 1.1 (16%), and 21.6 ps (8%). Using KCl instead of KI, the solvation time constants are nearly the same, 0.36 ps (74%), 1.2 ps (16%), and 23.4 ps (10%). These results clearly show that the cation has a significant effect, but the anion is negligible in the supramolecular solvation dynamics. The solvation dynamics of acetonitrile probed by small dye molecules (Coumarin 152 and 153) was reported to be less than 1 ps [35, 183]. The solvation dynamics observed here are different from that probed by simple dye molecules and contain considerably longer relaxation processes, which must be related to the local solvent structures. In acetonitrile, if the supramolecule is in the stretched form, the solvent arrangements around the indole moiety would be similar for both the supramolecule and tryptophan. The resulting solvation dynamics for both molecules would be on the same time scales. The observed long solvation dynamics in the supramolecule indicates a folded structure of **BI18C6** [Fig. 22b], not stretched, and certain solvent molecules are ordered and sandwiched between the marcoring and the indole ring. The $CH_3CN$ has a dipole moment of 3.92 D, which is much larger than that of $H_2O$ (1.85 D) and $CH_3OH$ (1.70 D). This large dipole moment and the rich negative-charge distributions of both the macroring (four oxygen and two nitrogen atoms) and the indole ring ($\pi$ electrons) can have large dipole–charge interactions, and as a result, the $CH_3CN$ molecules can be well oriented in a mostly parallel-type configuration with the linear C-C-N bonds parallel to two rings. Such parallel-aligned acetonitrile molecules have been observed recently in pure solution by X-ray diffraction [184]. Thus, the sandwiched solvent molecules are relatively rigid and have direct contacts with both indole rings. The solvent response to the sudden change of the indole dipole moment by electronic excitation is expected to be slow. The ordered local solvent structure results in enhancement of the local polarization, leading to a larger Stokes shift, which is consistent with the fluorescence emission in Fig. 23. Thus, solvent acetonitrile induces a folding of supramolecule **BI18C6** (Fig. 22b).

With the addition of KI, the slow component of 206 ps becomes much faster in about 20 ps. This observation suggests that the well-aligned local solvent structure is mostly rearranged. The interactions of the alkali cations with crown ethers such as 18-crown-6 have been studied very well, and in all the cases the cation is sitting at the center of the crown ring as a result of cation–oxygen interactions. With the addition of the side indole rings, the strong cation–$\pi$ interactions between $K^+$ and the two indole sidearms that fold down on the top and bottom of the crown ring have been observed [175]. The bonding energy of the alkali metal cations with indole is more than 30 kcal/mol [185]. Because of the steric effects and the strong electrostatic interactions, most solvent molecules originally sandwiched between the indole and crown ether rings are squeezed out. In this case, the well-aligned solvent structure around the macroring no longer exists. On the other side of the indole ring, the indole interacts with mostly bulk acetonitrile molecules, which results in much faster solvation dynamics; this finding is similar to the results probed by dye molecules. The exclusion of the ordered solvent molecules leads to a decrease of the local polarization and a less Stokes shift; the emission shifts back to the blue side by 10 nm. The observed 20-ps solvation dynamics but with a small 8% amplitude indicates that certain acetonitrile molecules still stick around the complex because of the charged cation and electron-rich crown ring. Experiments on another supramolecule (*N*-(2-(3-indolyl)ethyl)aza-18-crown-6) with only one indole sidearm gave similar results as for **BI18C6** with and without $K^+$; this result consistent with the symmetric folded structure and indicates that the indole ring mainly "senses" the solvent structure at one side of the macroring. We also measured anisotropy dynamics with and without encapsulation of $K^+$ in $CH_3CN$ (Fig. 25b). For **BI18C6** in $CH_3CN$, the long component is fitted with a time constant of 19 ps with an amplitude of 0.11. The 19 ps is the complete rotation relaxation time of **BI18C6** in $CH_3CN$ and no noticeable local wobbling motion of the indole ring was observed, indicating a relatively rigid structure, consistent with the folded compact structure. With the addition of KCl, the time constant slightly increases to 24 ps (not shown), which probably indicates a slight increase of the local viscosity by adding the salt. Changing from KCl to KI, the time constant becomes 31 ps and this increase is caused by the proximity of the heavy iodine anion to the supramolecule. The X-ray study revealed a hydrogen bonding between $I^-$ and the nitrogen of the indole ring [178]. However, if the **BI18C6** in $CH_3CN$ adopts a stretched form, this form would have a different molecular volume from the folded structure. Using the Stokes–Einstein–Debye hydrodynamics theory [186], the complete rotation relaxation of the stretched structure takes a longer time than the folded form at least by a factor of three but we observed a similar time. This observation suggests a similar structure, which is consistent with the solvation studies.

The dynamic process of the cation recognition by the supramolecule crown ether is governed by free energy evolution of total enthalpy and entropy. The desolvation of the cation solvent shell and destruction of the local solvent structure of the host supramolecule are unfavorable in energy, but they are compensated by the large energy stabilization through the two strong cation–$\pi$ (indole) and four cation–oxygen (crown ring) interactions. The solvation dynamics around the $K^+$ cation is at most as slow as tens of picoseconds [187]. Clearly, in acetonitrile the deconstruction of local solvent structures is the rate-limiting step to reach the final recognition, whereas in methanol both desolvation processes of the cation and the marcoring are critical to molecular recognition because they occur on the similar time scale of 10–20 ps. The strong electrostatic interactions of the cation with the electron-rich indole $\pi$-rings and oxygen atoms are the dominant recognition force and control the entire recognition process.

## IV. SURFACE HYDRATION VERSUS PROTEIN SOLVATION

The observed slow solvation above in tens of picoseconds at several protein surfaces [45, 46, 70, 71] is significant and biologically relevant. However, some concerns were raised regarding the origin of the slow relaxation. With some molecular dynamics simulations by others [188], the slow component was attributed to the protein solvation of nearby polar/charged residues, which does not result from hydrating water motions. In this section, we review two proteins [72, 73] and each one has only one tryptophan residue at the surface. We systematically mutated the neighboring polar or charged residues within about 7 Å around the probe to examine the contributions of protein solvation and surface-water hydration.

### A. The Enzyme Staphylococcus Nuclease

Figure 26 shows the X-ray structure of Staphylococcus nuclease (SNase) consisting of three $\alpha$-helices and a five-stranded $\beta$-barrel, with a total of 149

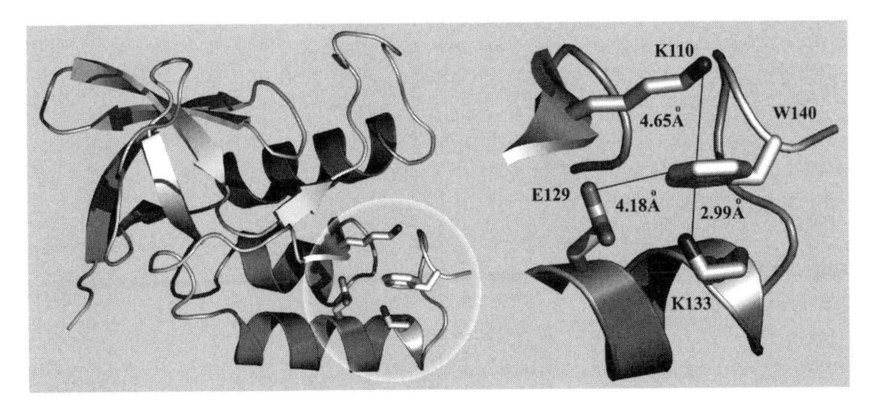


**Figure 26.** Left: X-ray crystallographic structure of wild-type SNase (PDB ID: 1SNO). The single tryptophan W140 is sandwiched between K110 and K133, with one edge exposed to the protein surface. Right: The local configuration around W140 with three charged residues (K110, K133, and E129) in close proximity (less than 5 Å).

amino acids [189]. The only single tryptophan residue (W140) has one edge exposed to the surface and inserts inside the protein to form a hydrophobic cluster at the C terminal, which was found to be essential to protein foldability, stability, and activity [190]. Three surface-charged residues, K110, K133, and E129 (Fig. 26), surround W140 within 7 Å. Using alanine scan, we replaced each charged residue with hydrophobic alanine one at a time by site-directed mutagenesis. We also mutated K110 into the polar residue cysteine. We measured the Stokes shifts, solvation dynamics, and local rigidity of the wild type and the mutants to determine the dependence of solvation energy and dynamics on charge distribution(s).

Figure 27 shows the steady-state fluorescence emission of the wild type and mutants, together with the transients gated at the red-side emission (360 nm). All emission peaks are around $332.5 \pm 0.5$ nm and no significant emission shift by the mutation of charged residues was observed. These results are striking and reveal that the three neighboring charged residues make a negligible contribution to solvation of the excited W140. These observations are consistent with the general trend that the maximum of tryptophan emission mainly depends on the location, i.e., the extent of its exposure to surface water, and not on its neighboring residues [53]. Thus, the observed Stokes shift cannot be caused by neighboring residues (with charges) solvation, which indicates relatively immobile charged side chains. The Stokes shift is dominantly due to hydration. From the X-ray structure (Fig. 26), W140 is sandwiched between K110 and K133, 4.65 Å and 2.99 Å, respectively, forming two cation–$\pi$ interactions. K133 is only 3.50 Å from E129, resulting in formation of a salt

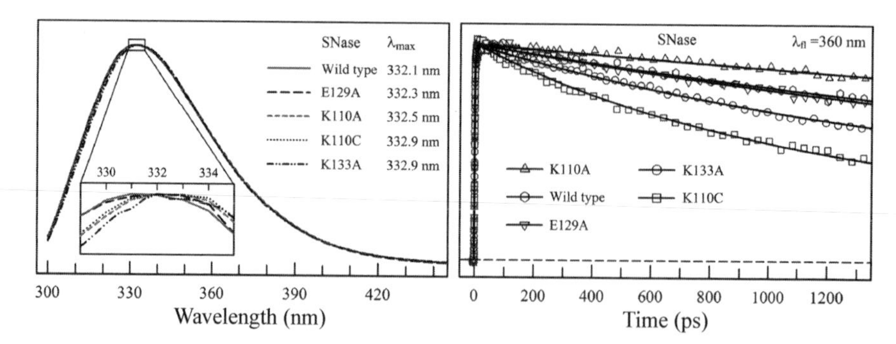


**Figure 27.** Emission spectra and lifetime transients of SNase. Left: Normalized steady-state fluorescence spectra of wild-type SNase and four mutants, E129A, K110A, K110C, and K133A. The mutant K110A has the strongest fluorescence intensity, and K110C has the weakest one because of the quenching by C110. Note that no significant emission shifts were induced by the mutation of charged residues. Right: Normalized fs-resolved fluorescence transients of W140 from wild-type SNase and all four mutants at the red-side emission (360 nm).

bridge. This unique structural motif with these strong electrostatic interactions is probably the origin of local protein rigidity around W140, which is consistent with the anisotropy dynamics, as discussed below.

Figure 28 shows the fs-resolved fluorescence transients of W140 in the wild type for several typical wavelengths, from the blue to red side, and for more than 10 gated emissions. Similarly, the overall decay dynamics is significantly slower than that of aqueous tryptophan in a similar buffer solution [49]. Clearly, the ultrafast decay components ($<1$ ps) observed in tryptophan solution were not observed at the blue side for the protein. The solvation components for all blue-side transients are well represented by a double-exponential decay with time constants that range from 4.9 to 12 ps and from 102 to 130 ps. For the red-side emission, the rise occurs in the range of 1.0 to 4.8 ps and is clearly present in all transients. For the four mutants of K133A, K110C, K110A, and E129A, besides the different lifetime emission contributions (Fig. 27), the transients showed similar solvation patterns from the blue to the red side. Among the four mutants, K110A has the shortest decay time for the two solvation components followed by E129A. The mutants of K133A and K110C have similar temporal behaviors as the wide type. The constructed solvation correlation functions are shown in Fig. 29. All solvation correlation functions can be represented by a double-exponential decay. For the wild type, the time scales are 5.1 ps with a 46% of the total amplitude and 153 ps (54%); for K110C, 4.2 ps (51%) and 149 ps (49%); for K133A, 3.9 ps (59%) and 157 ps (41%); for E129A, 3.5 ps (60%) and 124 ps (40%); for K110A, 3.1 ps (77%) and 96 ps (23%). Overall, all four mutants show faster temporal behaviors than the wild type, and all the

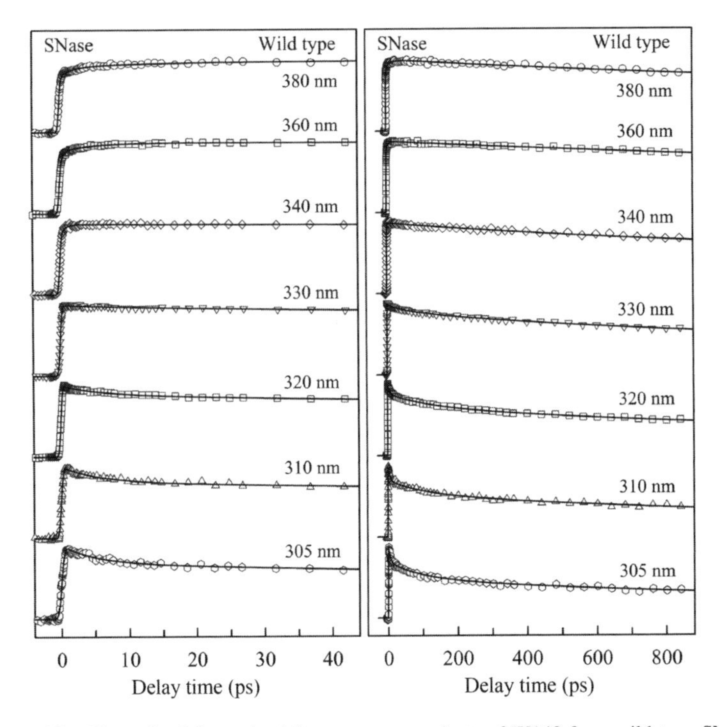


**Figure 28.** Normalized fs-resolved fluorescence transients of W140 from wild-type SNase on short (left) and long (right) time scales with a series of gated fluorescence emissions.

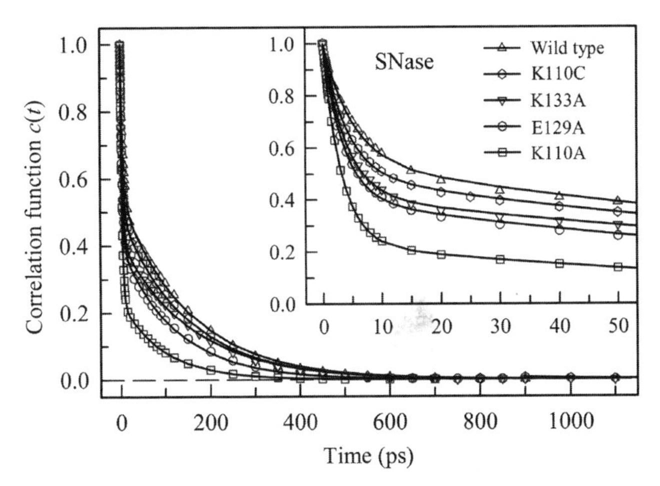


**Figure 29.** Hydration correlation functions $c(t)$ probed by tryptophan W140 for wild-type SNase and four mutants, K110C, K133A, E129A, and K110A. The inset shows the correlation functions in the short time range. Note the similarity of the time scales and the difference in amplitudes.

second long solvation times are within $\sim$100–150 ps. Note that no ultrafast solvation dynamics in less than 1 ps were observed for the proteins studied here.

The constructed solvation correlation function is the response of the local environment around W140 to the sudden dipole change from the ground state to the excited state. In principle, the response results from both surrounding water molecules and neighboring protein polar/charged residues (and protein peptide bonds). If the second long-time solvation *only* came from protein sidechain motions [188], in this case, from three neighboring charged residues, we should observe significantly different solvation time scales and amplitudes of the second solvation components. However, for the mutants K133A and K110C, we observed a similar long relaxation time (149 and 157 ps) as the wild type (153 ps). Overall, the obtained solvation dynamics for the four mutants become faster with decrease in the local charge distribution around W140, consistent with MD simulations that water has longer residence times around charged residues than near hydrophobic side chains [31, 32]. The observed long time scale of $\sim$100–150 ps is also consistent with the observation in melittin tetramer and human serum albumin (**E** and **F** isomers); in both cases [70, 71], the probe tryptophan has the similar emission maxima ($\sim$332 nm) and charge surroundings as in SNase. Thus, the obtained solvation correlation functions essentially reflect the local hydration dynamics at the protein surfaces.

For SNase, W140 has one edge exposed to the surface. Because of the local strong electrostatic interactions of the cation-$\pi$ stacking and salt bridge and the tight packing of the hydrophobic cluster near the C terminal, the local structure is more rigid and W140 is less mobile. We studied the fs-resolved rotational dynamics of W140 by the measurement of anisotropy changes with time for all the proteins. We observed that all anisotropy dynamics dominantly decay in nanoseconds with a small decay component at the early time. The long nanosecond dynamics represents the whole protein tumbling motion. The early small decays, on the time scales from 62 to 464 ps, result from the local wobbling motions with semiangles of 11°–19° [191]. The time scales of 62–464 ps do not directly correlate with those of solvation dynamics (Fig. 29). All these findings suggest a relatively small local fluctuation; the protein structure around W140 does not undergo large conformation changes on the time window of our measurements ($\sim$1.3 ns). These results from anisotropy dynamics are consistent with those obtained above from site-selected mutagenesis. The conclusion is indeed consistent with the results of the steady-state fluorescence emission maxima (*i.e.*, the dynamic Stokes shifts of the wild type and four mutant proteins are independent of three charged residues K110, K133, and E129). They, however, are contrary to the findings of recent MD simulations, which indicate that for this protein (SNase) K110 and K133 effects combine to make a very large contribution to the total Stokes shift [151].

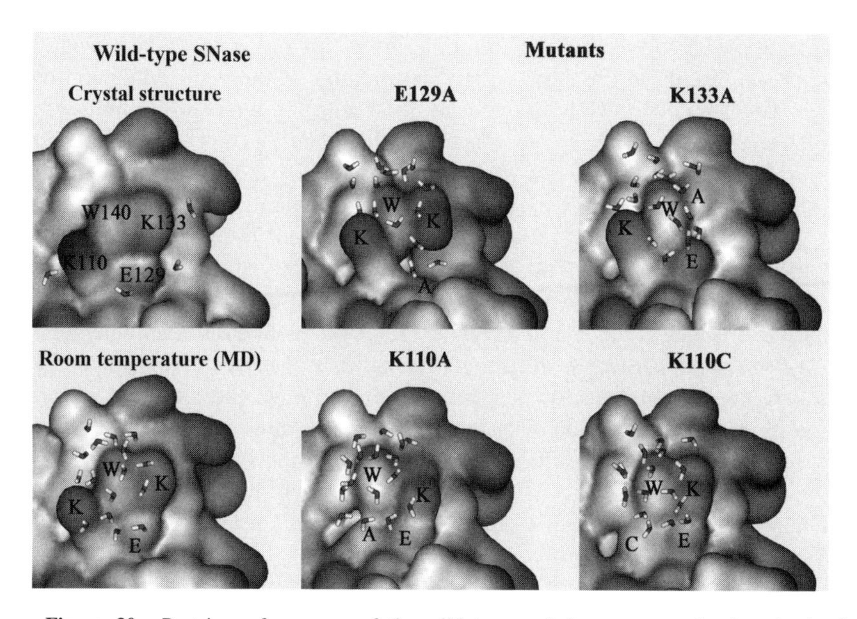


**Figure 30.** Protein surface maps of the wild type and four mutants showing the local topography and neighboring protein residues around W140. The light to dark colors represents increasing charges. Four water molecules stick to three surface-charged residues (K110, E129, and K133) near the probe W140 in the X-ray structure. The water molecules within 5 Å from the O atom of $H_2O$ to the indole ring are shown from our 2-ns single-trajectory MD simulations at 295 K in aqueous solution. Note the similar number of water molecules and different protein structures.

Thus, the studies of mutations, steady-state fluorescence, anisotropy measurements, and biphasic behavior of dynamic Stokes shifts exclude charged sidechain solvation as a significant contributor. Figure 30 shows a series of surface maps around W140 for the wild type and four mutants with the local surface topography and neighboring residues. The X-ray structure at 1.7 Å resolution shows around W140 four surface-water molecules sticking to three charged residues and one water molecule buried inside the protein. Within 5 Å from O atom of $H_2O$ to indole moiety (including H), there are about 17 water molecules around the ground state W140 from our recent 2-ns MD simulations. We did not observe significant changes of total water molecules for the different mutants, which implies nearly similar polar water environments; this finding consistent with our observation of the same Stokes shifts for all mutants and the wild type.

When tryptophan is excited to the $^1L_a$ state, the local equilibrium is shifted and the system is in a nonequilibrium state. Our observed hydration dynamics, with two distinct time scales, reflect the temporal evolution of two types of motions from the initial nonequilibrium configuration to new equilibrated state around the excited tryptophan. The initial hydration dynamics, which occur in 3–5 ps, represent the librational/rotational motions of these surrounding water molecules, slowing down by a factor of 3–5 compared with similar bulk-water motions around tryptophan, which occurs in less than 1 ps. We clearly observed a correlation of the initial fast time scale and amplitude with the local charge distribution. For example, we observed the fastest hydration dynamics in 3.1 ps for the mutant K110A with the largest amplitude (77%); for K110C, we obtained a time scale of 4.2 ps with 51%; for the wild type (K110), we have the initial dynamics of 5.1 ps with 46%. Thus, from the hydrophobic (A110) to the polar (C110) and to the charged residue (K110), we observed a lengthening of the time scale, consistent with stronger interactions of local water with charges. Furthermore, we observed a decrease, not increase, in the amplitude, which suggests that the charged/polar residue does not directly contribute to the solvation energy. These results reveal that the initial relaxation is from the local surface-water motions, and the observed variations in hydration dynamics by the mutations reflect the alternation of the local landscape and the change of the neighboring chemical identity (Fig. 30).

The observed slow water dynamics in 100–150 ps represent the long-time collective motions, which reflects a dynamically structural change of the local water molecules. The time scale to reach the new structural configuration depends on the local protein–water interactions. After the sudden change of the dipole moment of W140 in the protein, besides initial librational/rotational motions, at least the following two local structural motions are expected: One is the alignment of the local water network to the new excited-state ($^1L_a$) dipole moment, and the other is the increase of water molecules around W140 because of the larger excited-state dipole moment ($\Delta\mu{\sim}5$ D). Because of the longer time scale (tens of ps), these overall structural changes must be coupled with the protein fluctuation. These fluctuations assist in structural rearrangement of surface water and its exchange with bulk water. As such, the connection of residence time to hydration is incomplete without knowledge of fluctuations [16]. Thus, the long-time hydration dynamics, which are defined here as the rate for water structure to reach the new equilibrated state of minimal free energy, is an integrated process determined by the local interactions with the protein (Fig. 30) and assisted by its fluctuations. The protein–water coupling motions enable surface-water molecules to make structural arrangements (solvation), but these small protein fluctuations themselves do not make direct contributions to total solvation.

## B. The Electron-Transfer Protein Human Thioredoxin

We examined another system, an electron-transfer protein human thioredoxin (hTrx). Figure 31 shows the X-ray structure of oxidized hTrx at 2.1 Å resolution [192, 193]. hTrx is a compact globular protein with a central core of five strands

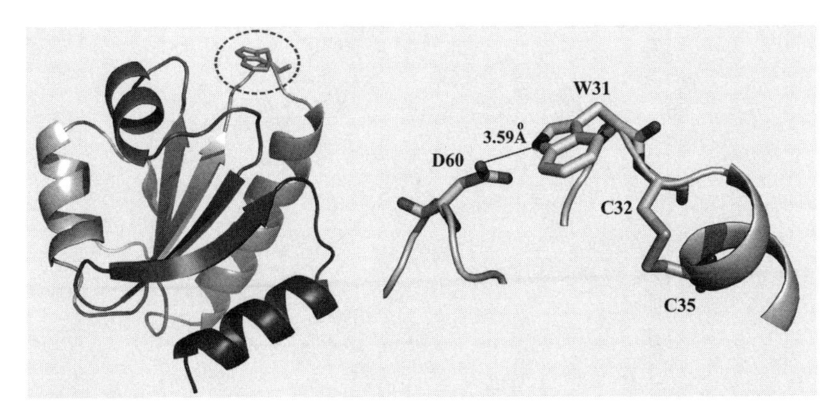


**Figure 31.** Structure of hTrx with W31 at the active site. Left: X-ray crystallographic structure of oxidized hTrx (C73S), which consists of five strands of $\beta$-pleated sheet and four $\alpha$-helices. The single tryptophan W31 (sticks in the broken circle) is lying down at the protein surface at the active site. Right: The local configuration in close proximity (within 7 Å) to W31 with one charged residue (D60) and a disulfide bridge (C32–C35).

of $\beta$-pleated sheet surrounded by four $\alpha$-helices. The active site of hTrx consists of a highly conserved sequence W31-C32-G33-P34-C35-K36 that forms a protruding part of the structure between the second $\beta$-strand and the second $\alpha$-helix. The enzyme contains only a single tryptophan (W31) that is lying down at the protein surface in the active site (Fig. 31). The reduced hTrx involves various biological functions for catalyzing dithiol–disulfide exchanges with substrate proteins.

The active site is fully exposed to water, and all chemical reactions that occur at the active site must couple with local water motions; this hydration process often modulates chemical changes through local water rearrangements. The active-site W31 is the ideal probe for such hydration studies. In SNase above, the probe W140 is partially inserted into the protein and heavily surrounded by three charged residues. For hTrx, except only one charged residue aspartate 60 (D60) at 3.59 Å (Fig. 31), the probe W31 is enclosed mostly by hydrophobic environment with its fluorescence emission at 339 nm (Fig. 32). We first engineered the distant cysteine 73 (C73) to serine (S) to avoid the protein dimer formation [194] and then mutated the charged residue D60 to hydrophobic glycine (G) and polar asparagine (N) to characterize the solvation dynamics of different mutants and determine whether any significant changes of solvation dynamics occur at the active site from mutations.

Similarly, we systematically measured the fs-resolved fluorescence transients from the blue to red side. The solvation components for all blue-side transients are well presented by a double-exponential decay with time constants that range

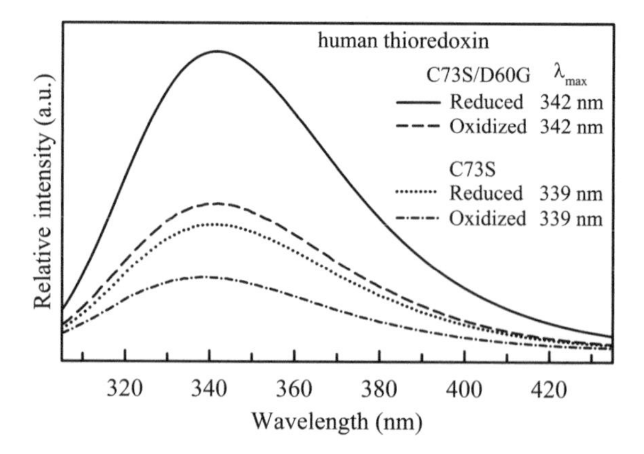


**Figure 32.** Steady-state fluorescence emission spectra of W31 in hTrx (C73S) and mutant D60G in reduced and oxidized states. For both mutants, the fluorescence intensity in reduced state is ~two fold of that in oxidized state. Also there is a ~two-fold increase in intensity as a result of the mutation of aspartate to glycine.

from 0.3 to 1 ps and from 2.6 to 12.6 ps. All transients at the red-side emission have a single rise component with time constants in the range of 0.7–1.3 ps. Besides the ultrafast solvation components, we also observed significant electron-transfer quenching dynamics by peptide bond and neighboring residues in hundreds of picoseconds [73]. Such significant quenching dynamics result in different apparent lifetimes and emission peaks. Thus, the difference in emission peaks of two mutants in Fig. 32 is not from changes in hydrophobicity but caused by the multiple emissions with different peaks for each mutant; this process is called decay associated emission spectra [78]. Nevertheless, because the solvation time scales are well separated from the quenching dynamics, we can construct solvation correlation functions for all mutants and the final results are shown in Fig. 33.

Strikingly, the solvation dynamics for all mutants are nearly the same. All correlation functions can be best described by a double exponential decay with time constants of 0.67 ps with 68% of the total amplitude and 13.2 ps (32%) for D60, 0.47 ps (67%) and 12.7 (33%) for D60G, and 0.53 ps (69%) and 10.8 ps (31%) for D60N. Relative to SNase above, the solvation dynamics are fast, which reflects the neighboring hydrophobic environment. We also measured the anisotropy dynamics and, as shown in the inset of Fig. 33, the local structure is very rigid in the time window of 800 ps. This observation is consistent with the inflexible turn (-T30W31-) in the transition from the second $\beta$-sheet and the second $\alpha$-helix (Fig. 31). Thus, the three mutants, with a charged, polar, or hydrophobic reside around the probe (Fig. 34) but with the similar time scales of

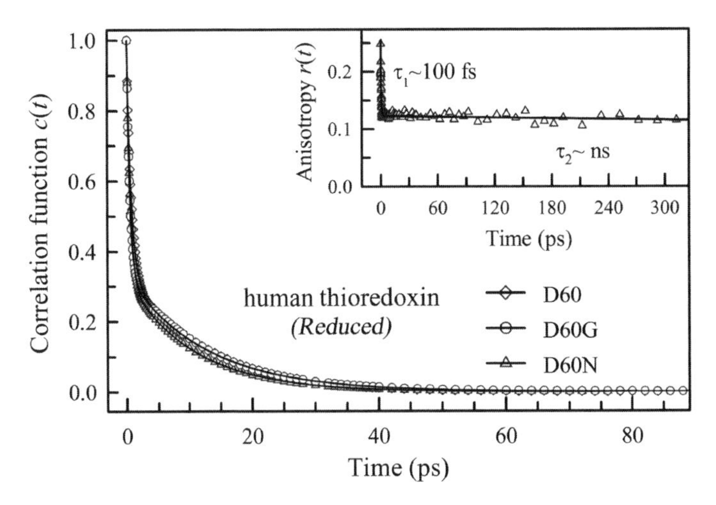


**Figure 33.** Hydration correlation functions $c(t)$ probed by W31 in hTrx and two mutants in reduced states. The three functions are strikingly similar, indicating a similar local solvent environment upon mutation. The inset shows the fluorescence anisotropy dynamics of W31. The nearly constant $r(t)$ reflects a very rigid local structure.

solvation dynamics in 0.47–0.67 and 10.8–13.2 ps, clearly show that the solvation contributions are dominantly from the local hydrating water molecules and not from the neighboring protein polar/charged residues. The protein side chain solvation is minor, and the dynamics of local water networks near the hydrophobic patches occur in 10–15 ps (Fig. 34).

In summary, various experimental data, especially with systematic mutations of Staphylococcus nuclease and human thioredoxin, support that the hydration dynamics makes the dominant contribution to the total solvation response

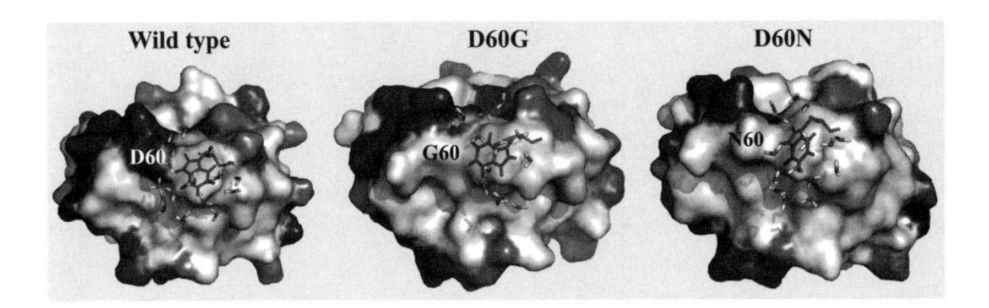


**Figure 34.** Surface map representations of the structure of wild-type hTrx and two mutants. The structures were generated with MD simulation. The white, light gray, and dark gray colors represent nonpolar, positively, and negatively charged residues, respectively, and W31 is shown as sticks. Note that local protein environment is very nonpolar.

obtained using surface tryptophan as a local optical probe. With extensive surface-Trp mutation work and molecular dynamics simulations below, we will continue to understand protein side-chain and surface water motions and thus reveal the intimate relationship of hydration dynamics and protein fluctuations around protein surfaces.

## V. GLOBAL MAPPING OF SURFACE HYDRATION AND PROTEIN FLUCTUATIONS

We have observed the dynamics of surface protein hydration on picosecond time scales with a robust biphasic distribution. To generalize the global heterogeneous hydration dynamics around protein surfaces, correlate the dynamics with protein local structures and chemical identities, and decipher the molecular mechanism of water–protein fluctuations, we summarize here our direct mapping of water motions around a globular protein, apomyoglobin (apoMb), in its two states, native and molten globular, using intrinsic tryptophan residue (W) as a local molecular probe to scan the surface by protein engineering [74]. Figure 35 shows myoglobin from sperm whale that consists of eight α-helices (A–H) with a total

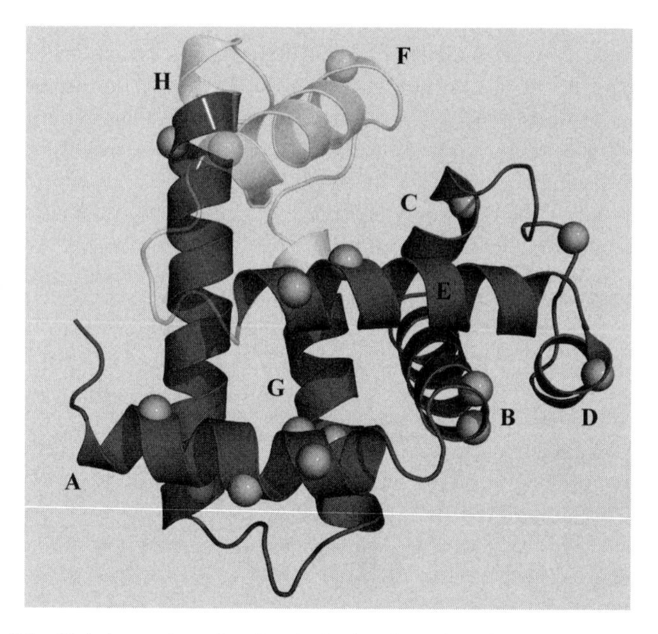


**Figure 35.** Global mapping of surface hydration dynamics of apoMb. Shown is the X-ray crystal structure of sperm whale myoglobin (PDB ID: 1MBD) in the holo form with eight helices A–H. In apo form, parts of the structure are melted and they are shown in transparent gray. The 16 balls indicate positions of mutation with tryptophan one at a time.

of 153 amino acids [195, 196]. Among 153 amino acid residues, two intrinsic tryptophan residues, W7 and W14, exist in the A helix. To achieve site-specific detection, we need only one Trp residue in each mutant. As the first step, we mutated tryptophan by tyrosine one at a time, and two mutant proteins W7Y (W14) and W14Y (W7) were obtained. To place a single tryptophan in other helices on the surface, we first mutated out both tryptophans (W7 and W14) in the A helix [197] and designed a third mutation at the desired position. All experiments were performed with apoMb by removal of the prosthetic heme group. We carefully designed more than 30 mutants and placed tryptophan one at a time along each helix at the protein surface. After we screened all mutant proteins with their structural content, stability, and excited-state lifetime of tryptophan, only 16 mutants are appropriate for mapping global hydration, as shown in Fig. 35.

Figure 36 shows the femtosecond-resolved fluorescence transients of A144W mutant in the native (pH = 6.0) and molten globule (pH = 4.0) states [198] for

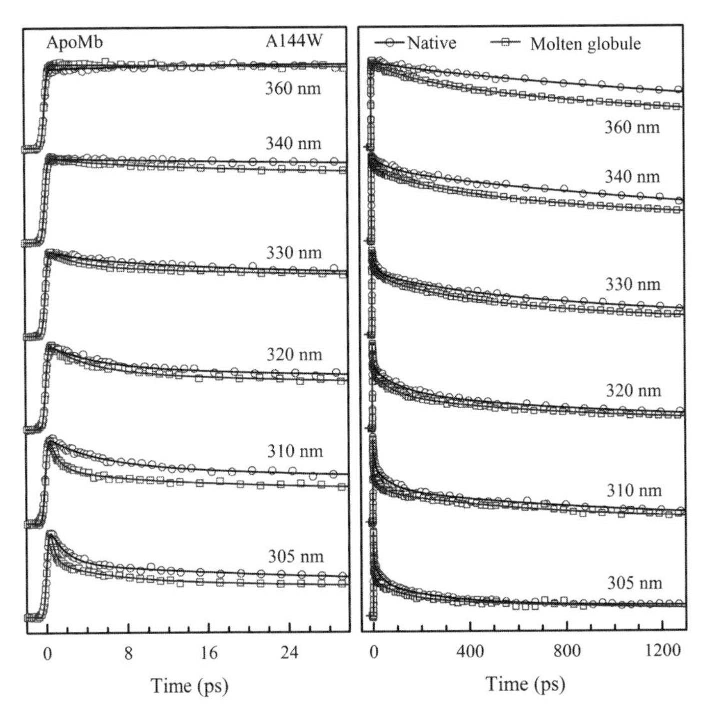


**Figure 36.** Fs-resolved fluorescence transients of mutant A144W for several gated emission wavelengths in the native and molten globule states in short (left) and long (right) time ranges. Note that all of the signals become faster in the molten globule state.

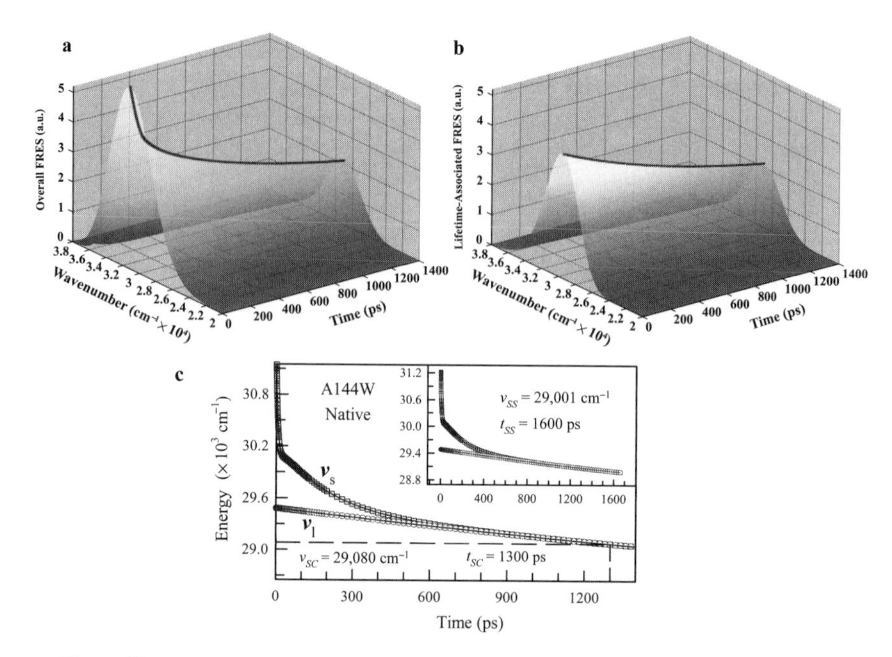


**Figure 37.** (a) 3-D representation of the evolution of overall FRES of mutant A144W in native state. (b) 3-D representation of the evolution of lifetime-associated FRES of mutant A144W in native state. (c) The temporal evolution of the emission peaks of overall ($\nu_s$) and lifetime-associated ($\nu_l$) FRES of A144W in native state. The inset shows $\nu_s$ and $\nu_l$ until both reach the steady-state emission peak $\nu_{ss}$.

several typical gated wavelengths. Similarly, the solvation components at the shorter wavelengths show ultrafast decays on picosecond time scales and gradually slow down toward the longer emissions. For all 16 mutant proteins and in two states, the fluorescence dynamics show a similar pattern of temporal behaviors with a double-exponential decay for solvation, but the time scales are different. With these transients, we constructed 3D presentation of FRES in Fig. 37 for both overall Stokes shifts and lifetime-associated shifts. For all mutants in two states, we obtained all 29 dynamic Stokes shifts with time, and the corresponding 29 solvation correlation functions with local protein structures are shown in Fig. 38. For all mutants, the correlation functions show a robust, biphasic distribution, and the dynamic Stokes shifts (total $\Delta E$, $\Delta E_1$, and $\Delta E_2$) and decay times ($\tau_1$ and $\tau_2$) of two relaxations are shown in Figs. 39 and 40. The first relaxation occurs ultrafast on the time scale of a few picoseconds ($\approx$ 1–8 ps), and the second dynamics takes a longer time in tens to hundreds of picoseconds ($\approx$ 20–200 ps).

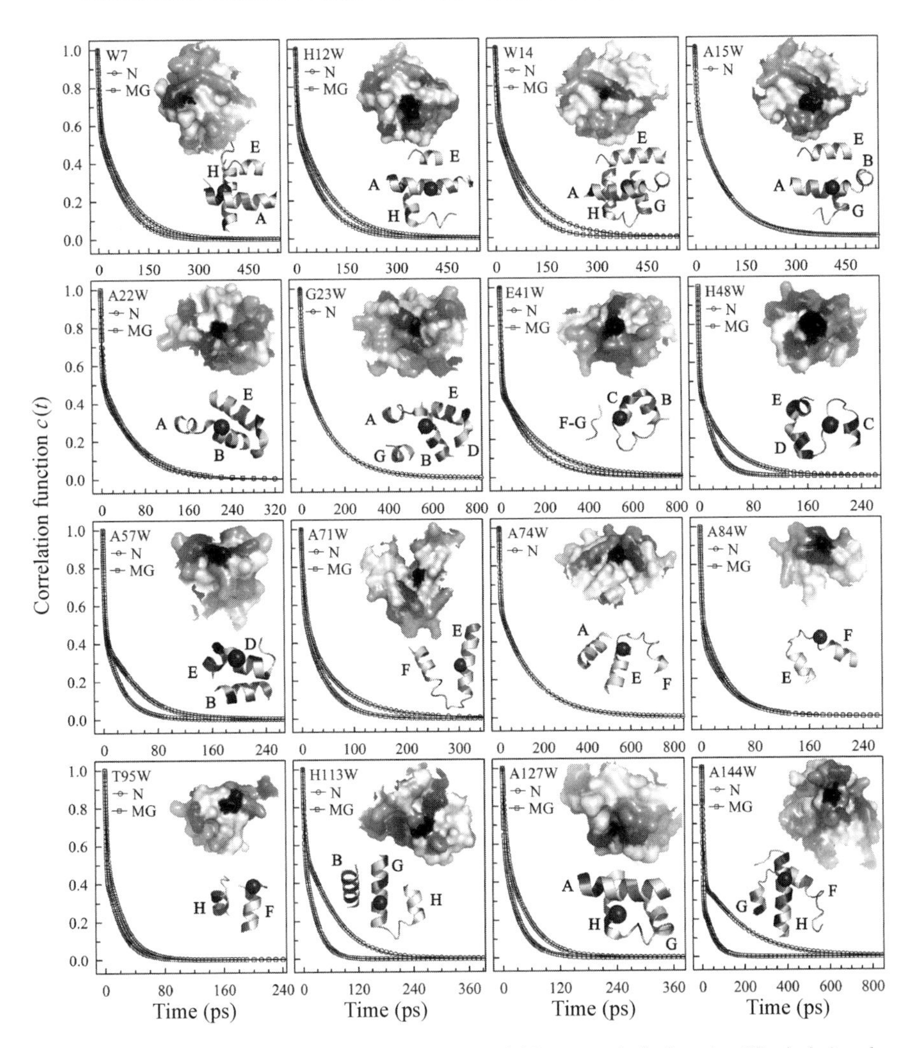


**Figure 38.** Hydration correlation functions $c(t)$ of 16 mutants in both native (N, circles) and molten globule (MG, squares) states. The solid lines are the best biexponential fit to $c(t)$. The insets show the local protein environment around sites of mutation both in surface map and ribbon representation. On surface maps, white, light gray, and dark gray colors represent nonpolar, positively, and negatively charged residues, respectively, and mutation sites are shown in black. On ribbon structures, mutation sites are indicated with black balls, and the A–H letters indicate identities of local helices.

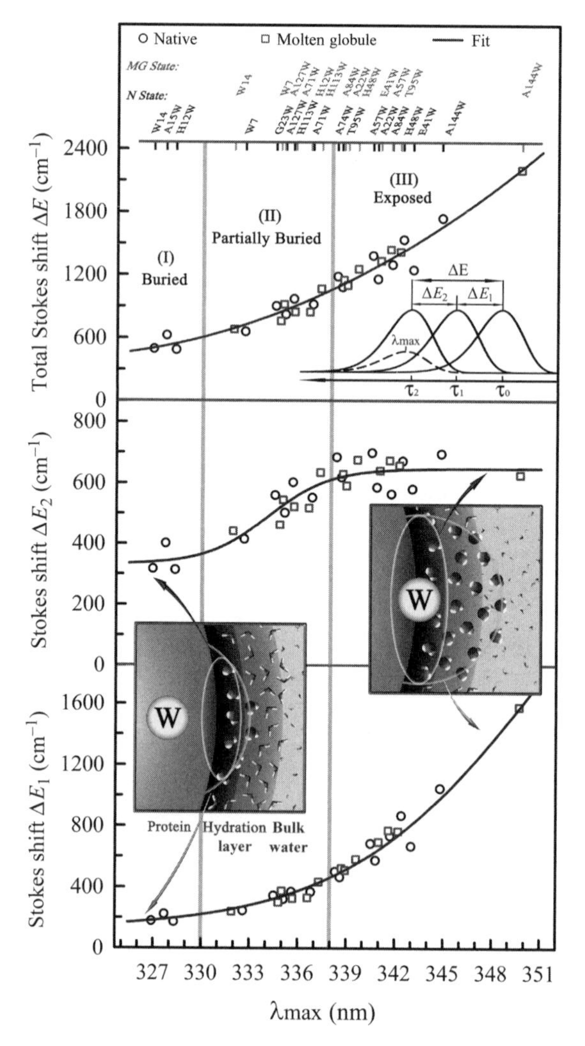


**Figure 39.** Dynamic Stokes shifts of all 16 mutants in two states. Circles and squares are the original data, and the black lines are the best fit. The names of mutants are shown on the top, and the ticks correspond to the data points. The inset in the top panel shows the physical meanings of the two Stokes shifts, $\Delta E_1$ and $\Delta E_2$. The insets in the lower two panels show different contributions of surface water to $\Delta E_1$ (big arcs, light arrow) and $\Delta E_2$ (small ellipse, dark arrow) when tryptophan is buried (left) or exposed (right). Water molecules in the big arcs are within ~10 Å around tryptophan, and water molecules in the small ellipse are those that directly interact with protein and probed by tryptophan.

## A. Dynamic Stokes Shifts and Two Distinct Water Motions

In Fig. 39, the three Stokes shifts show distinct relationships with tryptophan's emission maxima and ascertain the dominance of solvation response from surface water hydration. The first component $\Delta E_1$ shows a monotonic increase

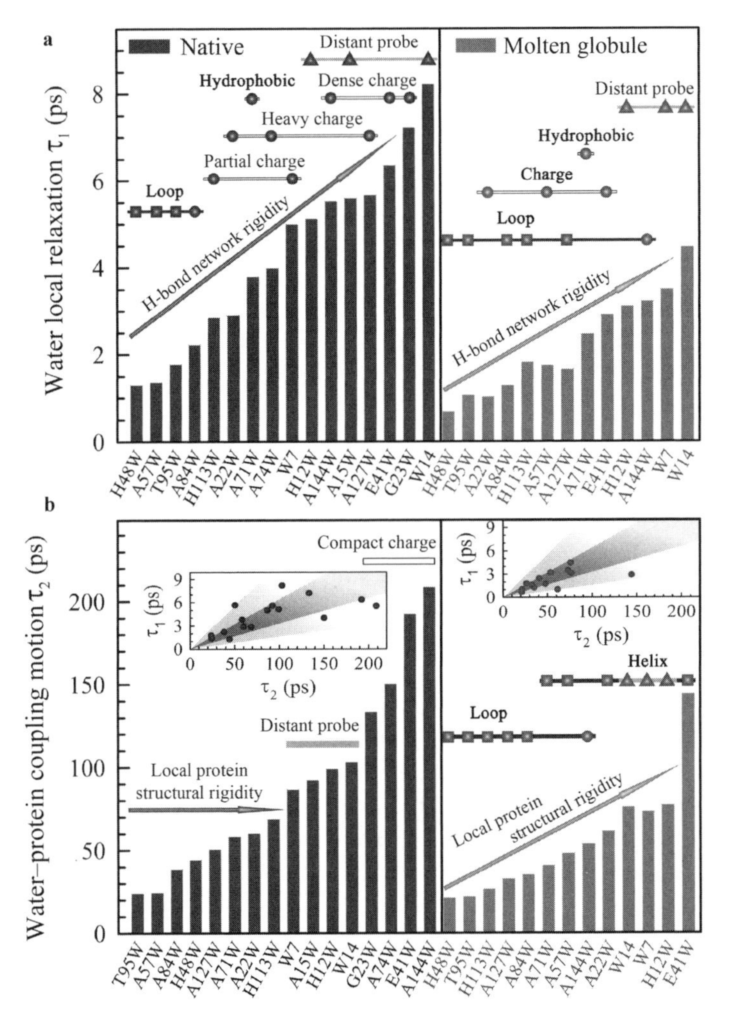


**Figure 40.** Time scales of two hydration dynamics component $\tau_1$ (top) and $\tau_2$ (bottom) of 16 mutants in native (right) and MG (left) states. In each panel, above the bars are the classification of mutation sites, which are divided according to their secondary structure (solid black bars), charge distribution (empty bars), and probe location (solid light gray bars). The beads correspond to the data below them. The insets in the lower two panels show the correlation between the two dynamics.

from 180 to 1600 cm$^{-1}$ when the probe gradually moves to the surface, which indicates that $\Delta E_1$ directly reflects the total probed water molecules through their local relaxation. The second component $\Delta E_2$ initially shows a gradual increase starting from 340 cm$^{-1}$ and reaches a plateau of ~650 cm$^{-1}$ at the emission maximum of 338 nm, which then becomes nearly independent of the probe surface locations as well as its local protein properties. This result is striking and must reflect a complete detection of the "intrinsic" solvation (~650 cm$^{-1}$) from large collective water motion in the hydration layer; this process of hydration water restructuring is assisted by local protein fluctuations. As demonstrated in our recent MD simulations below [199] and shown in *Staphylococcus* nuclease and human thioredoxin above, the small protein fluctuations themselves on the time scale of labile hydration do not make significant contributions to total solvation. The three regions (I, II, and III), which we divided according to the increasing trend of $\Delta E_2$, are exactly correlated to the well-known tryptophan locations with emission peaks in the protein [53] (i.e., from the buried below 330 nm, to the partially buried from 330 to 338 nm, and to the fully exposed at longer than 338 nm). For all these locations, based on different local protein properties, the maximum deviation along the general trend is $\pm$125 cm$^{-1}$ for $\Delta E_1$ and $\pm$75 cm$^{-1}$ for $\Delta E_2$.

### B. Time Scales, Correlations, and Water–Protein Fluctuations

The Stokes shift is the result of an integration of hydration processes but has no direct information on the dynamics. For example, both mutants of A84W and E41W have the similar Stokes shifts, but, as shown in Fig. 40, the dynamics are drastically different: 2.2 and 38 ps for the former and 6.4 and 192 ps for the latter. For $\tau_1$, the ultrafast dynamics represent fundamental motions of local water molecules, mainly, libration and hindered rotation. Compared with bulk water [49], the time scale of 1–8 ps is one order of magnitude longer, revealing different water structures and H-bond rigidity around the protein surface, which highly relates to local protein properties and directly reflects water–protein electrostatic interactions. According to the X-ray structure [196], we classified 16 mutants in the native state into several surface patterns such as the loop region and various charge patches (Fig. 40A). Each class covers a certain time regime. For example, the fastest motion in 1–2 ps (H48W, A57W, and T95W) is from the water molecules around the flexible loops, and these water molecules are less structured. The slow relaxation in 6–7 ps (G23W and E41W) is caused by the dense charged residues on the surface [Fig. 38], and the water molecules are tightly clustered. The water motion at the hydrophobic area near a concave geometry (A71W) takes about 4 ps. Also, along the A helix, the probes (H12W, A15W, and W14), buried inside the proteins, report even slower dynamics, which indicates the main probing of bound water molecules near the protein. For each class, the dynamics are correlated with the emission maxima (i.e., the longer emission the shorter $\tau_1$), which is caused by probing more and more mobile water molecules when tryptophan moves to the protein surface. These results show a strong correlation of the initial water dynamics with the local protein structures, chemical identities of residues, and probe locations. For the similar probe exposure, the time scales over 1–8 ps directly reflect local H-bond network rigidity on the protein surface.

The second hydration dynamics ($\tau_2$) in the layer represents subsequent water network rearrangements after the initial fast relaxation. On tens to hundreds of picoseconds, the protein fluctuates and the hydration water undergoes dynamic exchange with bulk water. Thus, such network restructuring couples with local protein fluctuations and convolutes with the exchange dynamics. We divided 16 mutants into three groups and Fig. 40b shows the correlation of the second water dynamics with local protein properties. Around four compact charge locations (G23W, A74W, E41W, and A144W), the relaxation takes the longest times in 133–209 ps because the water molecules are highly attracted by the local dense charges, as shown in Fig. 38 for A144W surrounded by five charged residues. Along the A helix, the four mutants (W7, A15W, H12W, and W14) surprisingly report a similar time scale of 87–103 ps, which reflects the "inherent" dynamics of the coupled water–protein relaxation around the A-helix region. Furthermore, for all other mutants, we observed a correlation between the dynamics and local protein structural rigidity from T95W to H113W (Fig. 40b). For example, the mutants in the loop regions such as T95W and A57W have the shortest relaxation time of 20 ps. With the increased rigidity of secondary and tertiary structures from A127W to H113W, the water–protein coupling motion gets longer and longer from 50 ps to 70 ps. All the second hydration dynamics have a wide correlation with the initial relaxation (the inset of Fig. 40b), which indicates that the network restructuring depends not only on H-bond network flexibility but also on local protein properties.

### C. Hydration Dynamics in Molten Globule State

In the molten globule state, NMR studies [198, 200, 201] have shown that the hydrophobic core is almost formed by three helices of A, G, and H. Except for the A57W mutant, the two hydration dynamics for all other mutants became faster, which suggests that both the H-bond water networks and the protein become more flexible and less structured. For some mutants shown in Fig. 38, the dynamics show the structural changes of hydration water and local protein from the native to molten globule states. For A144W, the dense charge distribution and well-structured H helix result in the slowest coupled water motion (209 ps); but in the molten globule state, the H-helix region near the C-terminal melts into a less-structured random coil [200]. The tryptophan is fully exposed to water (349.7 nm emission peak), and the relaxation becomes much faster. With dense charges around, it takes 53 ps. For W14 mutant on A-helix, the probe is buried, inside the protein, and we observed the similar dynamics in the two states, which indicates that the hydrophobic core around the probe has been formed at pH4. This result is consistent with the NMR results [198, 200]. For the H113W mutant, the partial B-helix structure melts down [210] and the local structure becomes flexible, which results in much faster coupled water–protein relaxation. For the A57W mutant, the hydration dynamics in the native state are much faster in 1.4 and 24 ps because of the flexible structure at the end of the D helix. However, in the molten globule state, the D helix extends longer [201], which results in a more rigid local structure around the probe and thus longer relaxation dynamics in 1.7 and 48 ps. Therefore, the probing of hydration dynamics can give structural information of proteins in the molten globule state.

## VI. MOLECULAR DYNAMICS SIMULATIONS

Direct MD simulations of the observed Stokes shifts and corresponding solvation time scales for several proteins were reported recently [188, 199, 202, 203]. Overall, significant discrepancies exist between simulation results and experimental observations, but some general features are promising. Here, we summarize one of our recent MD studies of W7 in apomyoglobin with linear response and direct nonequilibrium calculations and highlight the critical findings, as well as point out extensive improvement required in theoretical model [199].

The simulations for the protein were conducted using a double precision version of the GROMACS package and GROMOS96 force field with the SPC/E water model. The non-bonded pair list was produced using a 9 Å cut-off. Long-range electrostatic interactions were handled using the smoothed particle mesh Ewald (SPME) algorithm [204, 205] with a real space cutoff length of 9 Å. The cutoff length for the Lennard-Jones potential was set at 14 Å. All bond lengths were constrained using the LINCS algorithm [206], which allows a 2-fs time step in the simulation. Periodic boundary conditions were implemented using a truncated triclinic box of side length 60 Å and solvated with 4537 water molecules. The Nose-Hoover thermostat [207–209] was used to maintain the system at 295 K. The initial configuration of myoglobin is taken from the crystal structure (PDB:1MBD). Site charges of the indole chromophore in the $S_0$ state came form the GROMOS96 force field. For the $L_a$ excited state of the indole ring, which is the fluorescing state of Trp in the protein, we modified the partial charge of the indole chromophore by applying the ab initio charge density differences calculated by Sobolewski and Domcke [63] to the ground state partial charges of the GROMOS96 force field. At each configuration of the system, the energy difference between the $L_a$ and $S_0$ state associated with portion $x$ of the system ($x =$ protein, water, total) is denoted by $\Delta E^x(t)$, which in our simplified model is the Coulomb interaction difference between the system with excited and ground state charges on the indole of W7, as well as a gas phase transition energy, which is irrelevant for the calculated Stokes shift. The nonequilibrium Stokes shift is the average $\langle \Delta E^x(t) - \Delta E^x(0) \rangle$ over trajectories that evolve on the $L_a$ state surface with initial conditions sampled from an equilibrium ensemble on the $S_0$ surface. The linear response approximation to the Stokes shift from component $x$ of the system is given by $\frac{1}{k_B T}[\langle \Delta E^x(t)\Delta E(0) \rangle - \langle \Delta E^x(0)\Delta E(0) \rangle]$ (28) where $k_B$ is Boltzmann's constant and $T$ is the temperature.

Myoglobin has been studied extensively using MD simulations [30, 32, 210–222]. Most of these studies involve trajectories of much less than 1 ns except a few simulations of length 0.7–1.1 ns [216], 1–2 ns [217], 1 ns [219], 80 ns [220], and 90 ns [221]. We sampled a trajectory of ground-state myoglobin extending over 30 ns after initial equilibration for 800 ps and found structural fluctuations occurring on time scales extending up to several nanoseconds. Figure 41, which is a plot of the energy differences between the excited and ground states, shows sudden changes (20 kJ/mol) of the indole–water and indole–protein interaction energies after 10 ns occurring in a complementary way, which indicates a structural transition. More examination of the structures before and after 10 ns revealed that the loop backbone between helices E and F with residues 77–86 underwent significant displacement. Most of the resulting energy change is contained in the interaction between loop

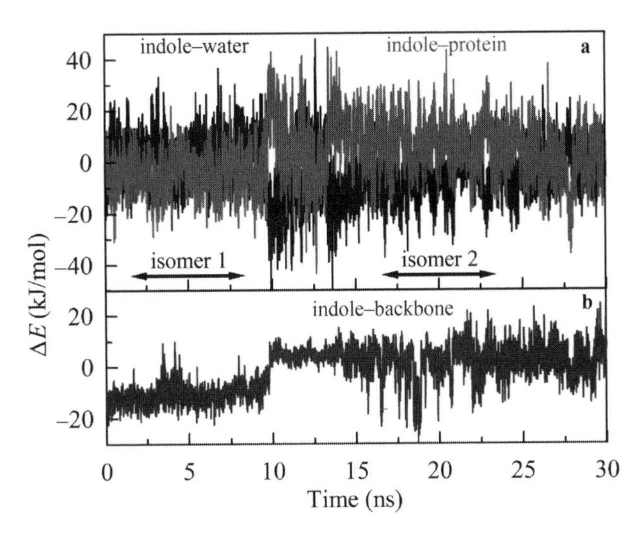


**Figure 41.** Time evolution of the energy differences between the excited and ground states for the indole (Trp7) with the protein and water in (a), and with the backbone of the loop (residues 79–86) in (b). A structural transition occurs at about 10 ns, and two structures, were used to perform MD simulations, which are defined as isomer 1 and isomer 2, respectively.

backbone and chromophore, as shown in Fig. 41b. The loop is closer to the indole during the first 10 ns. After the transition, the loop residues fluctuate more, and the helix F is more disordered near the loop. We refer to the structure of the first 10 ns of the trajectory as isomer 1, and the remainder of the trajectory as isomer 2 (Fig. 41a). This transition between isomer 1 and isomer 2 is probably a dynamic hopping between local substates [223, 224].

The relaxation dynamics (W7 in Fig. 38) is the response of the environment around Trp7 to its sudden shift in charge distribution from the ground state to the excited state. Under this perturbation, the response can result from both the surrounding water molecules and the protein. We separately calculated the linear-response correlation functions of indole–water, indole–protein, and the sum of the two. The results for isomer 1, relative to the time-zero values, are shown in Fig. 42a. The linear response correlation function is accumulated from a 6-ns interval indicated in Fig. 41a during which the protein was clearly in the isomer 1 substate. All three correlation functions show a significant ultrafast component: 63% for the total response, 50% for indole–water, and nearly 100% for indole–protein. A fit to the total correlation function beyond the ultrafast inertial decrease requires two exponential decays: 1.4 ps (3.6 kJ/mol) and 23 ps (2.0 kJ/mol). Despite the 6-ns simulation window for isomer 1, the 23-ps long component is not well determined on account of the noise apparent in the linear response correlation function (Fig. 42a) between 30 and 140 ps. The slow dynamics are mainly observed in the indole–water relaxation and the overall indole–protein interactions apparently make nearly no contributions to the slowest relaxation component.

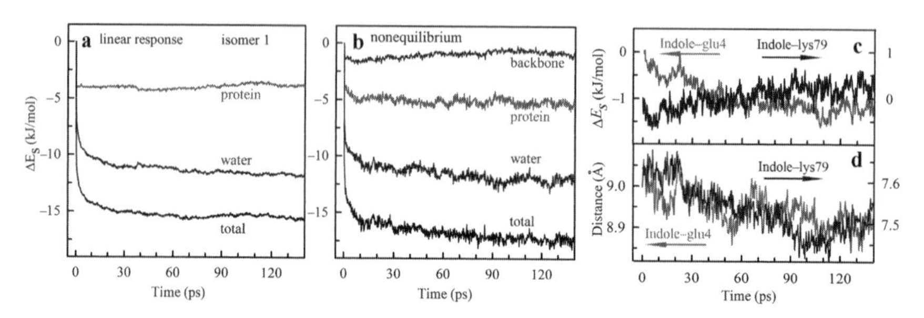


**Figure 42.** Solvation dynamics from MD simulations for isomer 1. (a) The linear-response calculated time-resolved Stokes shifts for indole–protein, indole–water, and their sum. (b) Direct nonequilibrium simulations of the time-resolved Stokes shifts for indole–water, indole–protein, and their sum. Note the lack of long-time component in indole–protein relaxation in both (a) and (b). Also shown is the indole–backbone interaction, whose magnitude diminishes at long times, complementary to the increase in water response. (c) Relaxation of two neighboring charged residues of a salt bridge, indole–lys79 and indole–glu4. The interaction energy changes from these two residues cancel each other. (d) The distance changes between the indole and two charged residues, which both move closer to the indole ring after photoexcitation.

We also performed direct nonequilibrium simulations following photoexcitation. The initial configurations were sampled from the same time interval used to generate the linear response correlation function (Fig. 41a). The results are shown in Fig. 42b for isomer 1 with the average of 360 trajectories. The nonequilibrium results up to 140 ps are similar to those from linear-response calculations in Fig. 42a. Indole–water interactions make the dominant contributions to the slow component of the total response, and the indole–protein contribution is minor after ~10 ps. The total relaxation dynamics can be fitted by a 18-fs inertial component (4.0 kJ/mol) and three exponentials: 80 fs (6.1 kJ/mol), 1.6 ps (4.9 kJ/mol), and 56 ps (2.7 kJ/mol), and the total Stokes shift is 17.7 kJ/mol (15.5 kJ/mol for linear response). Because the loop is flexible near the indole, we examined the indole-backbone interaction and found a rise in energy with time (Fig. 42b), coupled to a complementary decrease in the interaction with water. We studied the two neighboring charged residues of a salt bridge, glu4 and lys79. We found both residues moved close to the indole ring (Fig. 42d), but the interaction energies nearly cancel each other (Fig. 42c), which make no *apparent* contribution to the indole–protein stabilization. The data in Fig. 42 clearly indicate that the water contribution dominates the long-time Stokes shift, yet both the neighboring water and protein rearrange their local configurations following photoexcitation. The three time scales reflect initial inertial dynamics (<20 fs), fast local reorientational and translational motions (a few picoseconds), and slow surface hydration coupled with protein motions (tens of picoseconds).

Similarly, we studied the relaxation dynamics for isomer 2, and the results are shown in Fig. 43, where both linear response and nonequilibrium

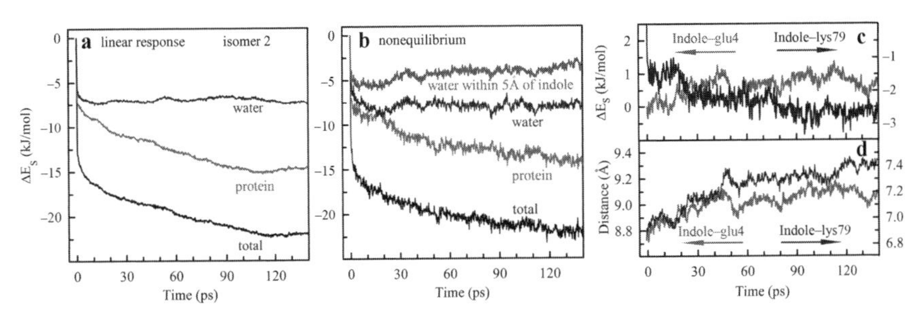


**Figure 43.** Solvation dynamics from MD simulations for isomer 2. (a) The linear-response calculated time-resolved Stokes shifts for indole–protein, indole–water, and their sum. (b) Direct nonequilibrium simulations of the time-resolved Stokes shifts for indole–water, indole–protein, and their sum. Note the lack of slow component in indole–water relaxation in both (a) and (b), which is opposite to isomer 1 in Fig. 42. Also shown is the indole–water (within 5 Å of indole) with coupled long-time negative solvation. (c) Relaxation between indole–lys79 and indole–glu4. The interaction energy changes from these two residues nearly cancel each other. (d) The distance changes between the indole and two charged residues, but both residues move away from the indole ring.

calculations, like isomer 1, exhibit similar dynamics within 140 ps (Fig. 43a and 43b). For linear response, the inertial response is 0.13 ps (12.5 kJ/mol) with another double-exponential decay of 2.2 ps (2.7 kJ/mol) and 67 ps (7.9 kJ/mol) and for direct nonequilibrium, the inertial response is 15 fs (2.9 kJ/mol) followed by another three-exponential decay of 75 fs (8.2 kJ/mol), 1.4 ps (4.7 kJ/mol), and 58 ps (6.7 kJ/mol). Besides the initial ultrafast components, the indole–protein contribution, in contrast to isomer 1, is dominant during 30–140 ps, and the indole–water contribution in this period is minor. The total Stokes shift calculated for isomer 2 is 22.5 kJ/mol (23.1 kJ/mol from linear response). Closer examination reveals underlying dynamics not apparent in the total Stokes shift or even in the water and protein components. Water molecules within 5Å from the indole (Fig. 43b) exhibit a significant solvation response up to the 140-ps limit of the nonequilibrium simulations in a direction that competes with indole–protein interactions. We also found that the loop backbone makes the major contribution to the indole–protein relaxation. Consistent with the structural flexibility observed in Fig. 41, the loop is more mobile. The neighboring charged residues glu4 and lys79 move away from the indole ring (Fig. 43d), but their interaction energies with indole also compensate each other (Fig. 43c). These results show that the flexible loop backbone dominates overall slow relaxation, and both the local protein and water molecules significantly rearrange, which leads to fortuitous cancellations of indole–water contributions in 140 ps.

The simulation results from both isomer 1 and isomer 2 show that the observed solvation dynamics around the Trp7 site can arise from strongly coupled neighboring water and protein relaxation. Judging by the time dependence of their separate contributions to the total response, the Stokes shift over tens of picoseconds can apparently result from either surface water or protein conformational relaxation for isomers 1 and 2, respectively. To elucidate the origin of these observed time scales, we performed frozen protein and frozen water simulations.

MD simulations with either protein or water constrained at the instant of photoexcitation were performed for both isomer 1 and isomer 2. For isomer 1, because surface water relaxation dominates the slow component of the total Stokes shift, in Fig. 44a we show the result of simulations of isomer 1 with an ensemble of frozen protein configurations to examine the role of protein fluctuations. Clearly the long component of indole–water interactions disappears when the protein is constrained. This result shows that without protein fluctuations, indole–water relaxation over tens of picoseconds does not occur. Thus, although surface hydrating water molecules seem to drive the global solvation and, from the dynamics of the protein and water contributions, are apparently responsible for the slowest component of the solvation Stokes shift for isomer 1 (Fig. 42), local protein fluctuations are still required to facilitate this rearrangement process. When the protein is frozen, the ultrafast

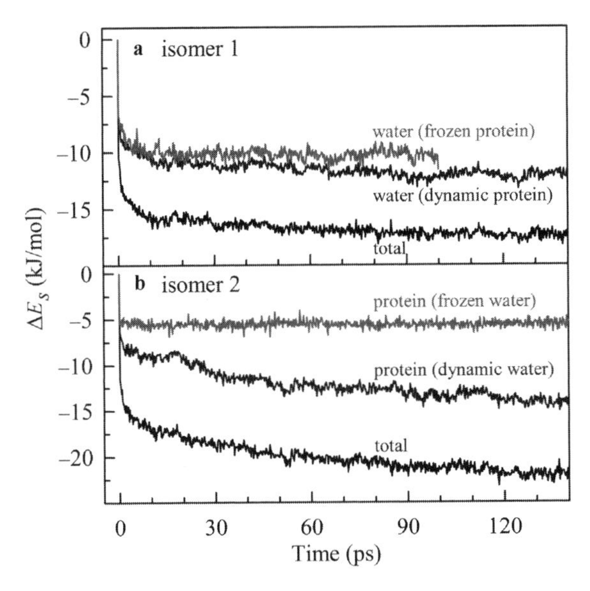


**Figure 44.** Solvation dynamics from constrained MD simulations. (a) Comparison of indole-water relaxation with and without frozen protein structure for isomer 1. The slow component of the water response nearly disappears, which indicates that slow water relaxation needs protein fluctuations. (b) Comparison of indole–protein relaxation with and without frozen water for isomer 2. Similarly, the slow component of the indole–protein disappears, which indicates that the protein relaxation also requires water fluctuations.

contributions (up to a few picoseconds) arise purely from inertial dynamics and surface water reorientations and translations. After these fast relaxation processes are complete, surface-water molecules under the rigid protein potential field are still in dynamical equilibrium with outside bulk water on a residence time scale of tens of picoseconds, just as they were when the protein was flexible, but this exchange dynamics does not lead to slow solvation dynamics without protein fluctuations. Protein fluctuations are effectively a part of surface hydration processes.

The fact that the indole–water relaxation is still present under the frozen protein with nearly the same amplitude (compare the water contribution in Fig. 42b and 44a) is one indication that the water response is not qualitatively modified by freezing the protein. The rigid potential field of the frozen protein somewhat limits rearrangements of the local water networks, but the difference is quantitative, not qualitative.

For isomer 2, protein interactions with the chromophore make the dominant contributions to the slow component of the Stokes shift. Therefore, in Fig. 44b we show simulation results under frozen water conditions. On the subpicosecond time scale, the protein can relax in the constrained water field, but the longer time indole–protein solvation disappears, which indicates that the protein cannot fluctuate under the rigid water potential field. The protein response following photoexcitation also requires water fluctuations. Thus, the slow solvation, protein fluctuations, and the dynamics of hydrating water are all intrinsically related.

The molecular mechanism of biphasic relaxation is shown in Fig. 45 based on MD simulations. The first 4 ps are dominated by local motions, like reorientation and translation of hydrating water molecules. Although water motions in the hydration layer are significantly longer than bulk water, water dynamics alone does not give rise to slow components in the Stokes shift that extend over tens of picoseconds. Instead, strongly coupled water–protein fluctuations are required. Constraining either the protein or water eliminates the slow component (Fig. 44). The interactions of surface water with protein fluctuations are an integral part of hydration dynamics, particularly at longer times. For the first time, we determined the actual time scale of coupled water–protein fluctuations around the site of Trp7 ($\sim$90 ps) from both an experimental and theoretical point of view. The dynamics of water–protein fluctuations is biologically significant and relevant, and it is shown now to occur on tens of picoseconds.

Some serious discrepancies however exist, as follows: (1) The total Stokes shifts from nonequilibrium calculations, 17.7 kJ/mole for isomer 1 and 22.5 kJ/mol for isomer 2, are significantly larger than the experimental result, 9.5 kJ/mol; (2) no ultrafast decay (or inertial motion) in less than 1 ps in the experiments

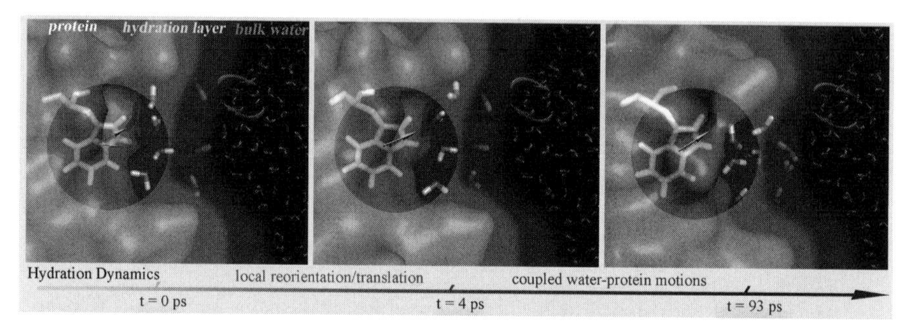


**Figure 45.** Illustration of the molecular mechanism of protein surface hydration with MD trajectories. Shown here is the Trp7 site with two hydration layers. Only some water molecules in the layers within 7 Å from the indole ring are shown for clarity. The twin arrows show a constant dynamical exchange of hydrating water molecules with bulk water. At time zero, the ground-state dipole of the indole is suddenly increased and reoriented by photoexcitation (left panel). The neighboring hydrating water molecules rapidly rotate and translate in a few picoseconds (middle panel). The next stage of the relaxation process is coupled water–protein motions over tens of picoseconds (right panel), which gives rise to the slow component of the Stokes shift. During this process, the initial water molecules may also be replaced by outside bulk water.

was observed, but the simulations gave more than 50% of the total energy on this time scale for both isomers; and (3) recent extensive simulations on other systems [188, 203] always show dominant protein solvation, which is contradictory to the experimental observations of water hydration [74]. These discrepancies could be resolved if the large inertial response observed in simulations were removed, either by improvement of the force field, such as by incorporation of electronic polarization [225, 226], or by inclusion of electronically nonadiabatic effects that might suppress the inertial regime. More work is clearly required to achieve quantitative agreement between theory and experiment.

## VII. CONCLUDING REMARKS AND FUTURE WORK

The robust observation of surface hydration dynamics on two time scales and a series of correlations with protein properties provides a molecular picture of water motions and their coupling with protein fluctuations in the layer, as shown in Fig. 46. The dynamic exchange of hydration layer water with outside bulk

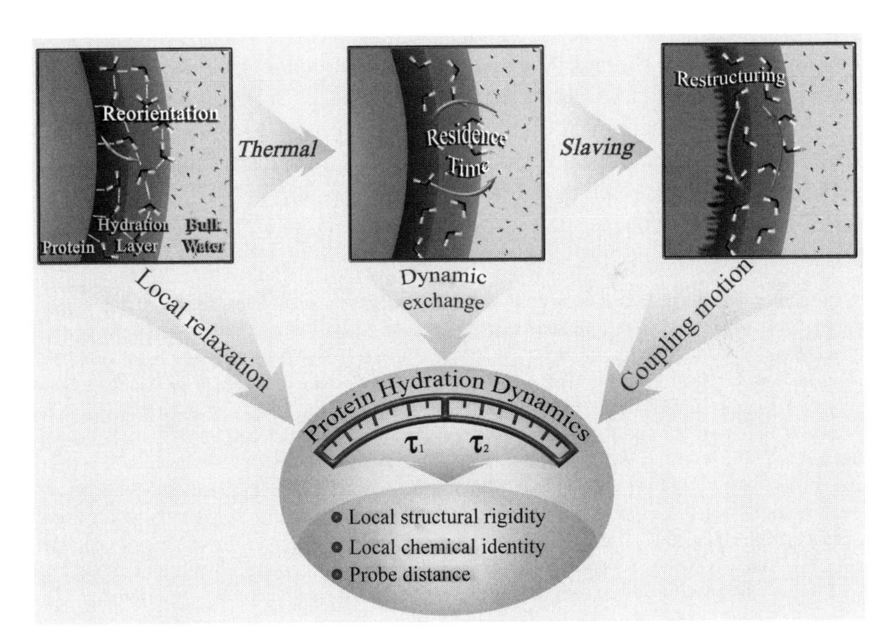


**Figure 46.** A unified molecular mechanism of protein hydration dynamics and coupled water–protein fluctuations. The initial ultrafast dynamics in a few picoseconds ($\tau_1$) represents local collective orientation or small translation motions, which mainly depend on local electrostatic interactions. On the longer time ($\tau_2$), the water networks undergo structural rearrangements in the layer, which are strongly coupled with both protein fluctuations and bulk-water dynamic exchange.

water occurs on the picosecond time scales (residence time), and such water thermal fluctuations "slave" local protein motions, especially for side chains. This process is similar to the $\beta$-relaxation in glasses [227, 228]. The hydration dynamics reported here represents the fluctuations of ordered water molecules inside the layer. The first ultrafast dynamics in a few picoseconds is directly from fundamental physical motions of water network at the local positions. Such cooperative H-bond motions are mainly determined by water–protein electrostatic interactions. On a longer time scale, the collective water network rearrangements in the layer strongly couple with protein fluctuations, all of which are facilitated by bulk-water dynamic exchange. Protein fluctuation-assisted interactions with surface water are an integral part of hydration dynamics, particularly at longer times. These dynamic processes are biologically relevant and significant and are now shown to occur on tens to hundreds of picoseconds. Thus, the dynamics of protein surface hydration uniquely bridge ultrafast bulk-water motions and slaved protein fluctuations and perform a biological functional role in maintaining protein's intact structure and flexibility, which lubricates its various recognitions and mediates a variety of catalytic reactions in enzymatic function.

Future work will naturally extend to study complex systems, such as hydration dynamics around different secondary-structure globular proteins; at interfaces of protein-DNA, RNA, or protein complexes; and at the active sites in enzymes. On the theoretical side, significant efforts are needed to solve the serious discrepancies of total solvation energy, ultrafast inertial motion, as well as protein flexibility and induced solvation.

## Acknowledgments

The author would like to thank Professors Ahmed Zewail (Caltech), Sherwin Singer (Ohio State Univ.), Ciriyam Jayaprakash (Ohio State Univ.), B. Montgomery Pettitt (Univ. of Houston), Steven Boxer (Stanford Univ.), Peter Rossky (UT Austin), James Hynes (Univ. of Colorado, Boulder), Wolfgang Doster (Tech. Univ. Munich), Douglas Tobias (UC Irvine), Patrik Callis (Montana State Univ.), Martin Gruebele (UIUC), Mark Berg (Univ. South Carolina), and David Leitner (Univ. of Nevada at Reno) for helpful discussions. Also, thanks to Professor Stephen Sligar (UIUC) for the protein plasmid and Professors David V.R. Sanders (Univ. of Saskatchewan), Wesley E. Stites (Univ. of Arkansas), George Gokel (Washington Univ.), and Marin Caffrey (Univ. Limerick) for certain experimental samples. Special thanks to Luyuan Zhang for the assistance in preparing this review, especially with all the figures; to Lijuan Wang for the critical role in protein biochemistry; and to all the group members for their work reported here. This work was supported in part by the National Science Foundation, the Packard Foundation Fellowship, and the Petroleum Research Fund.

## References

1. P. Ball, *Chem. Rev.* **108**, 74 (2008).
2. M. Chaplin, *Nat. Rev. Mol. Cell Biol.* **7**, 861 (2006).
3. K. Wuthrich, *Angew. Chem. Int. Ed.* **42**, 3340 (2003).

4. C. Mattos, *Trends Biochem. Sci.* **27**, 203 (2002).
5. J. L. Finney, *Faraday Discuss.* **103**, 1 (1996).
6. R. G. Bryant, *Annu. Rev. Biophys. Biomol. Struct.* **25**, 29 (1996).
7. S. N. Timasheff, *Annu. Rev. Biophys. Biomol. Struct.* **22**, 67 (1993).
8. M. M. Teeter, *Annu. Rev. Biophys. Biophys. Chem.* **20**, 577 (1991).
9. S. K. Pal and A. H. Zewail, *Chem. Rev.* **104**, 2099 (2004).
10. Y. Levy and J. N. Onuchic, *Annu. Rev. Biophys. Biomol. Struct.* **35**, 389 (2006).
11. J. C. Rasaiah, S. Garde, and G. Hummer, *Annu. Rev. Phys. Chem.* **59**, 713 (2008).
12. B. Bagchi, *Chem. Rev.* **105**, 3197 (2005).
13. R. Pethig, in *Protein Solvent Interactions*, R. B. Gregory, ed., Marcel Dekker, New York, 1995, p. 265.
14. L. D. Barron, L. Hecht, and G. Wilson, *Biochemistry* **36**, 13143 (1997).
15. V. Makarov, B. M. Pettitt, and M. Feig, *Acc. Chem. Res.* **35**, 376 (2002).
16. S. K. Pal, J. Peon, B. Bagchi, and A. H. Zewail, *J. Phys. Chem. B* **106**, 12376 (2002).
17. K. Bhattacharyya, *Chem. Commun.*, 2848 (2008).
18. E. H. Grant, R. J. Sheppard, and G. P. South, *Dielectric Behaviour of Biological Molecules in Solution.* Clarendon, Oxford, U.K., 1978.
19. E. H. Grant, *Bioelectromagnetics* **3**, 17 (1982).
20. F. T. Burling, W. I. Weis, K. M. Flaherty, and A. T. Brunger, *Science* **271**, 72 (1996).
21. M. Billeter, *Prog. Nucl. Magn. Reson. Spectrosc.* **27**, 635 (1995).
22. J. L. Finney, in *Water, a Comprehensive Treatise*, F. Franks, ed., Vol. 6, Plenum, New York, 1979, p. 47.
23. X. Cheng and B. P. Schoenborn, *J. Mol. Biol.* **220**, 381 (1991).
24. G. Otting, E. Liepinsh, and K. Wuthrich, *Science* **254**, 974 (1991).
25. G. Otting, *Prog. Nucl. Magn. Reson. Spectrosc.* **31**, 259 (1997).
26. V. P. Denisov, B. H. Jonsson, and B. Halle, *Nat. Struct. Biol.* **6**, 253 (1999).
27. K. Wuthrich, M. Billeter, P. Guntert, P. Luginbuhl, R. Riek, and G. Wider, *Faraday Discuss.* **103**, 245 (1996).
28. V. P. Denisov and B. Halle, *Faraday Discuss.* **103**, 227 (1996).
29. P. Ahlstrom, O. Teleman, and B. Jonsson, *J. Am. Chem. Soc.* **110**, 4198 (1988).
30. P. J. Steinbach and B. R. Brooks, *Proc. Nat. Acad. Sci. U.S.A.* **90**, 9135 (1993).
31. A. R. Bizzarri and S. Cannistraro, *J. Phys. Chem. B* **106**, 6617 (2002).
32. V. A. Makarov, B. K. Andrews, P. E. Smith, and B. M. Pettitt, *Biophys. J.* **79**, 2966 (2000).
33. W. Jarzeba, G. C. Walker, A. E. Johnson, M. A. Kahlow, and P. F. Barbara, *J. Phys. Chem.* **92**, 7039 (1988).
34. R. Jimenez, G. R. Fleming, P. V. Kumar, and M. Maroncelli, *Nature* **369**, 471 (1994).
35. M. L. Horng, J. A. Gardecki, A. Papazyan, and M. Maroncelli, *J. Phys. Chem.* **99**, 17311 (1995).
36. M. F. Kropman and H. J. Bakker, *Science* **291**, 2118 (2001).
37. J. S. Bashkin, G. McLendon, S. Mukamel, and J. Marohn, *J. Phys. Chem.* **94**, 4757 (1990).
38. D. W. Pierce and S. G. Boxer, *J. Phys. Chem.* **96**, 5560 (1992).
39. R. R. Riter, M. D. Edington, and W. F. Beck, *J. Phys. Chem.* **100**, 14198 (1996).
40. X. J. Jordanides, M. J. Lang, X. Y. Song, and G. R. Fleming, *J. Phys. Chem. B* **103**, 7995 (1999).
41. P. Changenet-Barret, C. T. Choma, E. F. Gooding, W. F. DeGrado, and R. M. Hochstrasser, *J. Phys. Chem. B* **104**, 9322 (2000).

42. D. Zhong, S. K. Pal, and A. H. Zewail, *ChemPhysChem* **2**, 219 (2001).
43. B. E. Cohen, T. B. McAnaney, E. S. Park, Y. N. Jan, S. G. Boxer, and L. Y. Jan, *Science* **296**, 1700 (2002).
44. J. Gilmore and R. H. McKenzie, *J. Phys. Chem. A* **112**, 2162 (2008).
45. S. K. Pal, J. Peon, and A. H. Zewail, *Proc. Nat. Acad. Sci. U.S.A.* **99**, 1763 (2002).
46. J. Peon, S. K. Pal, and A. H. Zewail, *Proc. Nat. Acad. Sci. U.S.A.* **99**, 10964 (2002).
47. P. Abbyad, W. Childs, X. Shi, and S. G. Boxer, *Proc. Nat. Acad. Sci. U.S.A.* **104**, 20189 (2007).
48. P. Abbyad, X. Shi, W. Childs, T. B. McAnaney, B. E. Cohen, and S. G. Boxer, *J. Phys. Chem. B* **111**, 8269 (2007).
49. W. Lu, J. Kim, W. Qiu, and D. Zhong, *Chem. Phys. Lett.* **388**, 120 (2004).
50. L. Zhang, Y.-T. Kao, W. Qiu, L. Wang, and D. Zhong, *J. Phys. Chem. B* **110**, 18097 (2006).
51. J. M. Beechem and L. Brand, *Annu. Rev. Biochem.* **54**, 43 (1985).
52. M. R. Eftink, in *Methods of Biochemical Analysis*, C. H. Suelter, ed., Vol. 35, Wiley, New York, 1991, p. 127.
53. J. R. Lakowicz, *Principles of Fluorescence Spectroscopy*, Springer, New York, 1999.
54. J. R. Lakowicz, ed., *Topics in Fluorescence Spectroscopy, Vol. 6: Protein Fluorescence*, Kluwer Academic, New York, 2000.
55. P. R. Callis, in *Methods in Enzymology*, Vol. 278, Academic Press, New York, 1997, p. 113.
56. J. R. Lakowicz, *Photochem. Photobiol.* **72**, 421 (2000).
57. A. G. Szabo and D. M. Rayner, *J. Am. Chem. Soc.* **102**, 554 (1980).
58. A. J. Ruggiero, D. C. Todd, and G. R. Fleming, *J. Am. Chem. Soc.* **112**, 1003 (1990).
59. J. W. Petrich, M. C. Chang, D. B. McDonald, and G. R. Fleming, *J. Am. Chem. Soc.* **105**, 3824 (1983).
60. R. P. Rava and T. G. Spiro, *J. Phys. Chem.* **89**, 1856 (1985).
61. D. Zhong, S. K. Pal, D. Zhang, S. I. Chan, and A. H. Zewail, *Proc. Nat. Acad. Sci. U.S.A.* **99**, 13 (2002).
62. X. H. Shen and J. R. Knutson, *J. Phys. Chem. B* **105**, 6260 (2001).
63. A. L. Sobolewski and W. Domcke, *Chem. Phys. Lett.* **315**, 293 (1999).
64. D. Zhong and A. H. Zewail, *Proc. Nat. Acad. Sci. U.S.A.* **98**, 11867 (2001).
65. C. Saxena, A. Sancar, and D. Zhong, *J. Phys. Chem. B* **108**, 18026 (2004).
66. D. Zhong, *Curr. Opin. Chem. Biol.* **11**, 174 (2007).
67. Y.-T. Kao, C. Saxena, L. Wang, A. Sancar, and D. Zhong, *Cell Biochem. Biophys.* **48**, 32 (2007).
68. Y.-T. Kao, C. Tan, S. H. Song, N. Ozturk, J. Li, L. Wang, A. Sancar, and D. Zhong, *J. Am. Chem. Soc.* **130**, 7695 (2008).
69. J. Stevens and D. Zhong, *J. Phys. Chem. B*, in preparation.
70. W. Qiu, L. Zhang, Y.-T. Kao, W. Lu, T. Li, J. Kim, G. M. Sollenberger, L. Wang, and D. Zhong, *J. Phys. Chem. B* **109**, 16901 (2005).
71. W. Qiu, L. Zhang, O. Okobiah, Y. Yang, L. Wang, D. Zhong, and A. H. Zewail, *J. Phys. Chem. B* **110**, 10540 (2006).
72. W. Qiu, Y.-T. Kao, L. Zhang, Y. Yang, L. Wang, W. E. Stites, D. Zhong, and A. H. Zewail, *Proc. Nat. Acad. Sci. U.S.A.* **103**, 13979 (2006).
73. W. Qiu, L. Wang, W. Lu, A. Boechler, D. A. R. Sanders, and D. Zhong, *Proc. Nat. Acad. Sci. U.S.A.* **104**, 5366 (2007).

74. L. Zhang, L. Wang, Y.-T. Kao, W. Qiu, Y. Yang, O. Okobiah, and D. Zhong, *Proc. Nat. Acad. Sci. U.S.A.* **104**, 18461 (2007).
75. Y. Chen and M. D. Barkley, *Biochemistry* **37**, 9976 (1998).
76. Y. Chen, B. Liu, H.-T. Yu, and M. D. Barkley, *J. Am. Chem. Soc.* **118**, 9271 (1996).
77. P. R. Callis and T. Liu, *J. Phys. Chem. B* **108**, 4248 (2004).
78. C. P. Pan, P. R. Callis, and M. D. Barkley, *J. Phys. Chem. B* **110**, 7009 (2006).
79. W. Qiu, T. Li, L. Zhang, Y. Yang, Y.-T. Kao, L. Wang, and D. Zhong, *Chem. Phys.* **350**, 154 (2008).
80. A. Jain, C. S. Purohit, S. Verma, and R. Sankararamakrishnan, *J. Phys. Chem. B* **111**, 8680 (2007).
81. R. Bhattacharyya, D. Pal, and P. Chakrabarti, *Protein Eng. Des. Sel.* **17**, 795 (2004).
82. T. R. Ioerger, C. Du, and D. S. Linthicum, *Mol. Immunol.* **36**, 373 (1999).
83. D. S. Larsen, K. Ohta, Q.-H. Xu, M. Cyrier, and G. R. Fleming, *J. Chem. Phys.* **114**, 8008 (2001).
84. K. Ohta, D. S. Larsen, M. Yang, and G. R. Fleming, *J. Chem. Phys.* **114**, 8020 (2001).
85. J.-C. Mialocq and T. Gustavsson, in *Fluorescence Spectroscopy: New Trends in Fluorescence Spectroscopy,* B. Valeur and J.-C. Brochon, eds., Springer-Verlag, Berlin, Germany, 2001, p. 61.
86. J. Kim, W. Lu, W. Qiu, L. Wang, M. Caffrey, and D. Zhong, *J. Phys. Chem. B* **110**, 21994 (2006).
87. W. Lu, W. Qiu, J. Kim, O. Okobiah, J. Hu, G. W. Gokel, and D. Zhong, *Chem. Phys. Lett.* **394**, 415 (2004).
88. D. A. Dougherty, *Science* **271**, 163 (1996).
89. E. A. Meyer, R. K. Castellano, and F. Diederich, *Angew. Chem. Int. Ed.* **42**, 1210 (2003).
90. A. A. Hassanali, T. Li, D. Zhong, and S. J. Singer, *J. Phys. Chem. B* **110**, 10497 (2006).
91. P. W. Fenimore, H. Frauenfelder, B. H. McMahon, and F. G. Parak, *Proc. Nat. Acad. Sci. U.S.A.* **99**, 16047 (2002).
92. Y.-T. Kao, C. Saxena, L. Wang, A. Sancar, and D. Zhong, *Proc. Nat. Acad. Sci. U.S.A.* **102**, 16128 (2005).
93. V. M. Rosenoer, M. A. Rothschild, and M. Oratz, eds., *Albumin Structure, Function, and Uses* Pergamon, Oxford, U.K., 1977.
94. C. Bertucci and E. Domenici, *Curr. Med. Chem.* **9**, 1463 (2002).
95. D. C. Carter and J. X. Ho, *Adv. Protein Chem.* **45**, 153 (1994).
96. M. Wardell, Z. M. Wang, J. X. Ho, J. Robert, F. Ruker, J. Ruble, and D. C. Carter, *Biochem. Biophys. Res. Commun.* **291**, 813 (2002).
97. J. A. Luetscher, *J. Am. Chem. Soc.* **61**, 2888 (1939).
98. J. F. Forster, in *The Plasma Proteins,* F. W. Putman, ed., Academic Press, New York, 1960.
99. J. Wilting, J. M. H. Kremer, A. P. Ijzerman, and S. G. Schulman, *Biochim. Biophys. Acta* **706**, 96 (1982).
100. V. R. Zurawski and J. F. Foster, *Biochemistry* **13**, 3465 (1974).
101. S. Curry, H. Mandelkow, P. Brick, and N. Franks, *Nat. Struct. Biol.* **5**, 827 (1998).
102. P. A. Zunszain, J. Ghuman, T. Komatsu, E. Tsuchida, and S. Curry, *BMC Struct. Biol.* **3**, 6 (2003).
103. E. S. Benson and B. E. Hallaway, *J. Biol. Chem.* **245**, 4144 (1970).
104. J. K. A. Kamal, L. Zhao, and A. H. Zewail, *Proc. Nat. Acad. Sci. U.S.A.* **101**, 13411 (2004).

105. A. Sytnik and I. Litvinyuk, *Proc. Nat. Acad. Sci. U.S.A.* **93**, 12959 (1996).
106. K. Flora, J. D. Brennan, G. A. Baker, M. A. Doody, and F. V. Bright, *Biophys. J.* **75**, 1084 (1998).
107. P. Marzola and E. Gratton, *J. Phys. Chem.* **95**, 9488 (1991).
108. J. K. A. Kamal and D. V. Behere, *J. Biol. Inorg. Chem.* **7**, 273 (2002).
109. M. K. Helms, C. E. Petersen, N. V. Bhagavan, and D. M. Jameson, *FEBS Lett.* **408**, 67 (1997).
110. A. Siemiarczuk, C. E. Petersen, C. E. Ha, J. Yang, and N. V. Bhagavan, *Cell Biochem. Biophys.* **40**, 115 (2004).
111. D. Zhong, A. Douhal, and A. H. Zewail, *Proc. Nat. Acad. Sci. U.S.A.* **97**, 14056 (2000).
112. N. E. Levinger, *Science* **298**, 1722 (2002).
113. N. Nandi, K. Bhattacharyya, and B. Bagchi, *Chem. Rev.* **100**, 2013 (2000).
114. K. Koga, G. T. Gao, H. Tanaka, and X. C. Zeng, *Nature* **412**, 802 (2001).
115. J. Fitter, R. E. Lechner, and N. A. Dencher, *J. Phys. Chem. B* **103**, 8036 (1999).
116. P. Jurkiewicz, J. Sykora, A. Olzynska, J. Humplickova, and M. Hof, *J. Fluoresc.* **15**, 883 (2005).
117. S. Y. Bhide and M. L. Berkowitz, *J. Chem. Phys.* **123** (2005).
118. P. G. de Gennes, *Rev. Mod. Phys.* **64**, 645 (1992).
119. Y. Kong and J. Ma, *Proc. Nat. Acad. Sci. U.S.A.* **98**, 14345 (2001).
120. R. Pomes and B. Roux, *Biophys. J.* **82**, 2304 (2002).
121. J. K. Lanyi, *J. Phys. Chem. B* **104**, 11441 (2000).
122. F. Garczarek and K. Gerwert, *Nature* **439**, 109 (2006).
123. E. B. Brauns, M. L. Madaras, R. S. Coleman, C. J. Murphy, and M. A. Berg, *J. Am. Chem. Soc.* **121**, 11644 (1999).
124. E. B. Brauns, M. L. Madaras, R. S. Coleman, C. J. Murphy, and M. A. Berg, *Phys. Rev. Lett.* **88**, 158101 (2002).
125. S. K. Pal, L. Zhao, and A. H. Zewail, *Proc. Nat. Acad. Sci. U.S.A.* **100**, 8113 (2003).
126. S. K. Pal, L. Zhao, T. Xia, and A. H. Zewail, *Proc. Nat. Acad. Sci. U.S.A.* **100**, 13746 (2003).
127. M. R. Harpham, B. M. Ladanyi, N. E. Levinger, and K. W. Herwig, *J. Chem. Phys.* **121**, 7855 (2004).
128. D. M. Willard, R. E. Riter, and N. E. Levinger, *J. Am. Chem. Soc.* **120**, 4151 (1998).
129. R. E. Riter, D. M. Willard, and N. E. Levinger, *J. Phys. Chem. B* **102**, 2705 (1998).
130. D. Pant, R. E. Riter, and N. E. Levinger, *J. Chem. Phys.* **109**, 9995 (1998).
131. K. Bhattacharyya and B. Bagchi, *J. Phys. Chem. A* **104**, 10603 (2000).
132. K. Bhattacharyya, *Acc. Chem. Res.* **36**, 95 (2003).
133. I. R. Piletic, H. S. Tan, and M. D. Fayer, *J. Phys. Chem. B* **109**, 21273 (2005).
134. H. S. Tan, I. R. Piletic, R. E. Riter, N. E. Levinger, and M. D. Fayer, *Phys. Rev. Lett.* **94**, 057405 (2005).
135. H. S. Tan, I. R. Piletic, and M. D. Fayer, *J. Chem. Phys.* **122**, 174501 (2005).
136. D. Pant and N. E. Levinger, *Chem. Phys. Lett.* **292**, 200 (1998).
137. D. Pant and N. E. Levinger, *J. Phys. Chem. B* **103**, 7846 (1999).
138. Y. Maniwa, H. Kataura, M. Abe, S. Suzuki, Y. Achiba, H. Kira, and K. Matsuda, *J. Phys. Soc. Jpn.* **71**, 2863 (2002).
139. J. Cheng, S. Pautot, D. A. Weitz, and X. S. Xie, *Proc. Natl. Acad. Sci. U.S.A* **100**, 9826 (2003).
140. D. S. Venables, K. Huang, and C. A. Schmuttenmaer, *J. Phys. Chem. B* **105**, 9132 (2001).
141. H. Shirota and K. Horie, *J. Phys. Chem. B* **103**, 1437 (1999).

142. A. I. Kolesnikov, J. M. Zanotti, C. K. Loong, P. Thiyagarajan, A. P. Moravsky, R. O. Loutfy, and C. J. Burnham, *Phys. Rev. Lett.* **93**, 035503 (2004).
143. E. M. Landau and J. P. Rosenbusch, *Proc. Nat. Acad. Sci. U.S.A.* **93**, 14532 (1996).
144. M. Caffrey, *Curr. Opin. Struct. Biol.* **10**, 486 (2000).
145. H. Luecke, B. Schobert, J. K. Lanyi, E. N. Spudich, and J. L. Spudich, *Science* **293**, 1499 (2001).
146. P. Raman, V. Cherezov, and M. Caffrey, *Cell. Mol. Life Sci.* **63**, 36 (2006).
147. J. Clogston, G. Craciun, D. J. Hart, and M. Caffrey, *J. Control. Release* **102**, 441 (2005).
148. J. Faeder and B. M. Ladanyi, *J. Phys. Chem. B* **104**, 1033 (2000).
149. J. Faeder and B. M. Ladanyi, *J. Phys. Chem. B* **105**, 11148 (2001).
150. E. G. Finer, *J. Chem. Soc., Faraday Trans.* **69**, 1590 (1973).
151. J. T. Vivian and P. R. Callis, *Biophys. J.* **80**, 2093 (2001).
152. K. Hristova and S. H. White, *Biophys. J.* **74**, 2419 (1998).
153. M. C. Wiener and S. H. White, *Biophys. J.* **61**, 434 (1992).
154. B. de Foresta, J. Gallay, J. Sopkova, P. Champeil, and M. Vincent, *Biophys. J.* **77**, 3071 (1999).
155. M. A. Wilson and A. Pohorille, *J. Am. Chem. Soc.* **116**, 1490 (1994).
156. H. Raghuraman and A. Chattopadhyay, *Biophys. J.* **87**, 2419 (2004).
157. H. Raghuraman and A. Chattopadhyay, *Langmuir* **19**, 10332 (2003).
158. A. Chattopadhyay and R. Rukmini, *FEBS Lett.* **335**, 341 (1993).
159. A. S. Ladokhin, S. Jayasinghe, and S. H. White, *Anal. Biochem.* **285**, 235 (2000).
160. C. D. Tran and G. S. Beddard, *Eur. J. Biophys.* **13**, 59 (1985).
161. E. John and F. Jahnig, *Biophys. J.* **54**, 817 (1988).
162. D. Roccatano, G. Colombo, M. Fioroni, and A. E. Mark, *Proc. Nat. Acad. Sci. U.S.A.* **99**, 12179 (2002).
163. J. H. Lin and A. Baumgaertner, *Biophys. J.* **78**, 1714 (2000).
164. M. Bachar and O. M. Becker, *J. Chem. Phys.* **111**, 8672 (1999).
165. Y. K. Cheng and P. J. Rossky, *Nature* **392**, 696 (1998).
166. Y. K. Cheng, W. S. Sheu, and P. J. Rossky, *Biophys. J.* **76**, 1734 (1999).
167. S. Ohki, E. Marcus, D. K. Sukumaran, and K. Arnold, *Biochim. Biophys. Acta* **1194**, 223 (1994).
168. S. Frey and L. K. Tamm, *Biophys. J.* **60**, 922 (1991).
169. C. E. Dempsey, *Biochim. Biophys. Acta* **1031**, 143 (1990).
170. J. M. Lehn, *Angew. Chem. Int. Ed.* **27**, 89 (1988).
171. D. J. Cram, *Angew. Chem. Int. Ed.* **27**, 1009 (1988).
172. C. J. Pedersen, *Angew. Chem. Int. Ed.* **27**, 1021 (1988).
173. G. W. Gokel, *Crown Ethers and Cryptands*, Royal Society of Chemistry, Cambridge, U.K., 1991.
174. J. A. Semlyen, ed., *Large Ring Molecules*, Wiley, Chichester, U.K., 1996.
175. G. W. Gokel, W. M. Leevy, and M. E. Weber, *Chem. Rev.* **104**, 2723 (2004).
176. R. M. Izatt, K. Pawlak, J. S. Bradshaw, and R. L. Bruening, *Chem. Rev.* **91**, 1721 (1991).
177. S. L. De Wall, E. S. Meadows, L. J. Barbour, and G. W. Gokel, *Proc. Nat. Acad. Sci. U.S.A.* **97**, 6271 (2000).
178. J. X. Hu, L. J. Barbour, and G. W. Gokel, *Proc. Nat. Acad. Sci. U.S.A.* **99**, 5121 (2002).
179. G. W. Gokel, L. J. Barbour, R. Ferdani, and J. X. Hu, *Acc. Chem. Res.* **35**, 878 (2002).

180. J. X. Hu, L. J. Barbour, and G. W. Gokel, *J. Am. Chem. Soc.* **124**, 10940 (2002).
181. S. L. De Wall, E. S. Meadows, L. J. Barbour, and G. W. Gokel, *J. Am. Chem. Soc.* **121**, 5613 (1999).
182. E. S. Meadows, S. L. De Wall, L. J. Barbour, and G. W. Gokel, *J. Am. Chem. Soc.* **123**, 3092 (2001).
183. W. Jarzeba, G. C. Walker, A. E. Johnson, and P. F. Barbara, *Chem. Phys.* **152**, 57 (1991).
184. T. Takamuku, M. Tabata, A. Yamaguchi, J. Nishimoto, M. Kumamoto, H. Wakita, and T. Yamaguchi, *J. Phys. Chem. B* **102**, 8880 (1998).
185. J. C. Ma, and D. A. Dougherty, *Chem. Rev.* **97**, 1303 (1997).
186. J. S. Baskin, M. Chachisvilis, M. Gupta, and A. H. Zewail, *J. Phys. Chem. A* **102**, 4158 (1998).
187. W. Lu and D. Zhong, *J. Phys. Chem. B*, in preparation.
188. L. Nilsson and B. Halle, *Proc. Nat. Acad. Sci. U.S.A.* **102**, 13867 (2005).
189. D. M. Truckses, J. R. Somoza, K. E. Prehoda, S. C. Miller, and J. L. Markley, *Protein Sci.* **5**, 1907 (1996).
190. S. Hirano, H. Kamikubo, Y. Yamazaki, and M. Kataoka, *Proteins* **58**, 271 (2005).
191. R. F. Steiner, in *Topics in Fluorescence Spectroscopy*, J. R. Lakowicz, ed., Plenum, New York, 1991, p. 1.
192. S. K. Katti, D. M. Lemaster, and H. Eklund, *J. Mol. Biol.* **212**, 167 (1990).
193. A. Weichsel, J. R. Gasdaska, G. Powis, and W. R. Montfort, *Structure* **4**, 735 (1996).
194. J. F. Andersen, D. A. R. Sanders, J. R. Gasdaska, A. Weichsel, G. Powis, and W. R. Montfort, *Biochemistry* **36**, 13979 (1997).
195. S. E. V. Phillips and B. P. Schoenborn, *Nature* **292**, 81 (1981).
196. T. Takano, *J. Mol. Biol.* **110**, 537 (1977).
197. I. Sirangelo, S. Tavassi, P. L. Martelli, R. Casadio, and G. Irace, *Eur. J. Biochem.* **267**, 3937 (2000).
198. F. M. Hughson, P. E. Wright, and R. L. Baldwin, *Science* **249**, 1544 (1990).
199. T. Li, A. A. P. Hassanali, Y.-T. Kao, D. Zhong, and S. J. Singer, *J. Am. Chem. Soc.* **129**, 3376 (2007).
200. D. Eliezer, J. Yao, H. J. Dyson, and P. E. Wright, *Nat. Struct. Biol.* **5**, 148 (1998).
201. D. Eliezer, J. Chung, H. J. Dyson, and P. E. Wright, *Biochemistry* **39**, 2894 (2000).
202. T. Li, A. A. Hassanali, and S. J. Singer, *J. Phys. Chem. B*, **112**, 16121 (2008).
203. A. A. Golosov and M. Karplus, *J. Phys. Chem. B* **111**, 1482 (2007).
204. T. Darden, D. York, and L. Pedersen, *J. Chem. Phys.* **98**, 10089 (1993).
205. U. Essmann, L. Perera, M. L. Berkowitz, T. Darden, H. Lee, and L. G. Pedersen, *J. Chem. Phys.* **103**, 8577 (1995).
206. B. Hess, H. Bekker, H. J. C. Berendsen, and J. Fraaije, *J. Comput. Chem.* **18**, 1463 (1997).
207. S. Nose, *Mol. Phys.* **52**, 255 (1984).
208. S. Nose, *J. Chem. Phys.* **81**, 511 (1984).
209. W. G. Hoover, *Phys. Rev. A* **31**, 1695 (1985).
210. R. M. Levy, R. P. Sheridan, J. W. Keepers, G. S. Dubey, S. Swaminathan, and M. Karplus, *Biophys. J.* **48**, 509 (1985).
211. E. R. Henry, W. A. Eaton, and R. M. Hochstrasser, *Proc. Nat. Acad. Sci. U.S.A.* **83**, 8982 (1986).
212. R. Elber and M. Karplus, *Science* **235**, 318 (1987).
213. C. L. Brooks, *J. Mol. Biol.* **227**, 375 (1992).

214. S. Furoiscorbin, J. C. Smith, and G. R. Kneller, *Proteins* **16**, 141 (1993).

215. J. B. Clarage, T. Romo, B. K. Andrews, B. M. Pettitt, and G. N. Phillips, *Proc. Nat. Acad. Sci. U.S.A.* **92**, 3288 (1995).

216. J. D. Hirst and C. L. Brooks, *Biochemistry* **34**, 7614 (1995).

217. T. Simonson and C. L. Brooks, *J. Am. Chem. Soc.* **118**, 8452 (1996).

218. C. Rovira and M. Parrinello, *Int. J. Quantum Chem.* **80**, 1172 (2000).

219. A. L. Tournier and J. C. Smith, *Phys. Rev. Lett.* **91**, 208106 (2003).

220. M. Aschi, C. Zazza, R. Spezia, C. Bossa, A. Di Nola, M. Paci, and A. Amadei, *J. Comput. Chem.* **25**, 974 (2004).

221. C. Bossa, M. Anselmi, D. Roccatano, A. Amadei, B. Vallone, M. Brunori, and A. Di Nola, *Biophys. J.* **86**, 3855 (2004).

222. W. Gu and B. P. Schoenborn, *Proteins* **22**, 20 (1995).

223. H. Frauenfelder, F. Parak, and R. D. Young, *Annu. Rev. Biophys. Biophys. Chem.* **17**, 451 (1988).

224. H. Frauenfelder, S. G. Sligar, and P. G. Wolynes, *Science* **254**, 1598 (1991).

225. J. S. Bader and B. J. Berne, *J. Chem. Phys.* **104**, 1293 (1996).

226. P. V. Kumar and M. Maroncelli, *J. Chem. Phys.* **103**, 3038 (1995).

227. P. W. Fenimore, H. Frauenfelder, B. H. McMahon, and R. D. Young, *Proc. Nat. Acad. Sci. U.S.A.* **101**, 14408 (2004).

228. H. Frauenfelder, P. W. Fenimore, G. Chen, and B. H. McMahon, *Proc. Nat. Acad. Sci. U.S.A.* **103**, 15469 (2006).

# AUTHOR INDEX

Numbers in parentheses are reference numbers and indicate that the author's work is referred to although his name is not mentioned in the text. Numbers in *italic* show the page on which the complete references are listed.



*Advances in Chemical Physics, Volume 143*, edited by Stuart A. Rice

# SUBJECT INDEX



*Advances in Chemical Physics, Volume 143*, edited by Stuart A. Rice